THE POWER OF CONTINUOUS DELIVERY IN DEVOPS

Reduce Difficulty By Increasing Frequency Of Your Product Releases

Series: "Continuous Everything"

Book 2

Edition 1

November, 2017

Juni Mukherjee

Owner, Continuity

Web: http://continuity.world

ABOUT THE AUTHOR

Juni is a thought citizen in the DevSecOps space. She has functioned as an engineer, architect, product owner and consultant in Silicon Valley. She is an author and her debut book was "Continuous Delivery Pipeline – Where Does It Choke?". She interacts with the throbbing DevSecOps community on Quora and has published articles on TechBeacon. Some of the conferences she has spoken at are Jenkins World, AllDayDevOps, DevNet Create, PrairieDevCon (PrDC) and Knowledge Discovery on Databases (KDD). She also participates as a panelist in Continuous Discussions (C9D9).

Her official name is Maitrayee, as in:

"**M**y **A**dorable **I**ntrovert, **T**ake **R**est **A**s **Y**ou **E**at **E**ggs".

Yes, she loves eggs but that's barely the point unless you have wondered (like she did) whether we can implement Continuous Delivery of eggs. She was born in June and her pet name is Juni since birth. She goes by Juni - nothing official about it. It wouldn't have been this breezy if she was born in, let's say, February, would it?

Juni owns a consulting firm "*Continuity*" that helps build resilient processes to power your business. "*Continuity*" partners with you on your digital transformation journey and helps take your business KPIs (Key Performance Indicators) to the next level. Web: http://continuity.world.

Table Of Contents

About The Author 2

Table Of Contents 3

Table Of Figures 6

Chapter 1 | Introduction 10
DevSecOps: The New DevOps? 11
Continuous Everything. And Everyone. 12
Acknowledgement 13

Chapter 2 | Continuous Delivery As A Domain 14
Domain Events 15
Domain Model For Continuous Delivery Pipeline 16
Domain Services 24
Pipeline Domain Model Fragmentation 26
NodeJS Pipeline Model 28
Java Pipeline Model 29
iOS Pipeline Model 31
Android Pipeline Model 32
Firmware and Embedded Pipeline Model 34
IoT Pipeline Model 35

Chapter 3 | Pipeline Domain Model Integrity 37
Bounded Context 38
Ubiquitous Language 38
Continuous Integration 39
Context Map 40
Shared Kernel 40
Customer/Supplier 41
Conformist 43
Open Host Service, Published Language 45
Anticorruption Layer 46
Big Ball Of Mud (And Tests) 47
Separate Ways 49

Chapter 4 | Continuous Analytics And Insights 51
Organizational KPIs Vs. Departmental KPIs 52

% Of Deployments To Dev Vs. Stage Vs. Prod 53
% Of Failures Per Test Type 54
% Of Defects Discovered Vs. Escaped 56
Business Value Delivered Per Sprint 57
Weighted Stability Index 58
Weighted Code Quality Index 59
Check-In-2-Go-Live With # Of Escaped Defects 61
Concept-2-Cash 62

CHAPTER 5 | A DEVSECOPS SEED BACKLOG 64
For Tech Leads 65
For Engineers 76
For Product Owners 89

CHAPTER 6 | THE TWELVE-FACTOR PIPELINE 92
i. Codebase 93
ii. Dependencies 94
iii. Configuration 96
iv. Backing Services 97
v. Build, Release, Run 98
vi. Processes 99
vii. Port Binding 101
viii. Concurrency 101
ix. Disposability 102
x. Dev/Prod Parity 103
xi. Logs 104
xii. Admin Processes 104

CHAPTER 7 | SEGREGATION/SEPARATION OF DUTIES 106
SoD In The US Government 107
SoD In The Enterprise 107
Current State Of DevSecOps And CD 108
Core Principles Of SoD 110
Modeling SoD In The Pipeline Domain Model 111
SoD Core Principle #1 111
SoD Core Principle #2 112
SoD Core Principle #3 114
SoD Core Principle #4 115

CHAPTER 8 | MYTH BUSTERS 117
Continuous Everything Is Overkill 118
Cannot Start Due To Manual Tests 118
More Environments Mean More Quality 120

VSM Has No Implications On CD 121
Teams Must Be Co-located...122
Engineers Can't Write Poems!..123
Engineering Owns Vision, Not Execution 125
*aaS Is Difficult To Embrace 126

CHAPTER 9 | RESOURCES 128
Websites 128
Books 128

TABLE OF FIGURES

This book has the following figures embedded. The figures are color-coded for ease of understanding.

Figure 1 - Domain Model For Continuous Delivery Pipeline 16
Figure 2 - Domain Services For Continuous Delivery Pipeline 24
Figure 3 - Fragmentation Of Pipeline Domain Model 27
Figure 4 - Firmware And Embedded Systems 34
Figure 5 – IoT Continuous Delivery of Eggs 35
Figure 6 - Customer/Supplier 42
Figure 7 - Conformist 44
Figure 8 - Anticorruption Layer 46
Figure 9 - Product Composition Anti-pattern 48
Figure 10 - Independently Deployable Artifact 49
Figure 11 - Key Performance Indicators 52
Figure 12 – Concept-2-Cash 62
Figure 13 - Test Architecture For Microservices 67
Figure 14 - Publish Subscribe Architectural Pattern 85
Figure 15 - Circuit-breaker Pattern 88
Figure 16 - Hand-off Anti-pattern 109

CAVEAT: The elaborate nature of some figures can cause them to get pixelated, both in electronic and print formats. Hence, the figures (*.png) have been made available on Continuity's website: http://continuity.world.

In the e-book (comes free with the paperback), you can click on each figure's caption to access its original copy situated in a Google Drive:
https://drive.google.com/drive/folders/0BzGPyPGE1d46SkYwZTZncDVvZDg?ths=true.

INDEX

A/B, 23, 76, 84
Amazon, 28, 29, 30, 31, 32, 33, 46, 47, 79, 94, 99
analytics, 69, 85, 124
Android, 25, 27, 32, 33, 41, 95
App Store, 32
Appium, 31, 32, 33, 95
architecture, 11, 21, 25, 47, 49, 67, 68, 90
artifact repository, 19, 47, 71, 79, 80
artifacts, 15, 19, 24, 28, 30, 31, 33, 34, 47, 49, 68, 71, 72, 73, 80, 88, 96, 98, 120, 121
audit, 25, 40, 72, 73, 74, 79, 82, 100
auditable, 45, 96, 98
automated, 12, 15, 19, 20, 21, 23, 40, 56, 65, 66, 70, 73, 74, 75, 79, 84, 87, 90, 95, 97, 103, 122
automation, 56, 57, 82, 119, 120
AWS, 24, 25, 26, 45, 46, 47, 71, 79, 88, 93, 94, 95, 96, 99
Azure, 24, 46, 71, 79, 88, 94, 96
BlazeMeter, 25, 29, 30, 71, 88, 95, 96
bounded context, 38, 40
Build, 17, 18, 19, 20, 22, 23, 28, 30, 31, 33, 42, 43, 44, 70, 98, 128
CD, 12, 89, 110, 121, 122
certified, 72, 80, 86, 120
Check-In-2-Go-Live, 58, 59, 61, 62, 65, 66, 70, 89, 91, 104, 110
Cloud, 25, 54, 76, 94, 95, 99, 101, 103, 120
Cloud Foundry, 24, 46, 71, 79, 88, 95, 96
Cobertura, 25, 28, 30, 41, 95
compatibility, 22, 49
compliance, 25, 82
component, 43, 44
components, 19, 47, 68, 80, 86
Concept-2-Cash, 62, 63, 66
configuration, 25, 45, 57, 64, 71, 80, 82, 83, 84, 88, 91, 93, 94, 96, 97, 101, 103, 116, 119, 120
Consumer, 41, 42, 43, 44
containers, 17, 20, 96, 99, 101, 103
Continuous, 1, 2, 10, 12, 13, 14, 15, 16, 21, 24, 27, 28, 34, 35, 36, 37, 38, 39, 41, 45, 47, 51, 54, 56, 57, 61, 63, 66, 68, 69, 70, 72, 73, 75, 76, 80, 81, 89, 91, 95, 96, 97, 98, 104, 106, 108, 110, 111, 118, 120, 121, 122, 125, 128
Continuous Delivery, 2, 10, 12, 14, 15, 16, 21, 24, 34, 35, 36, 37, 39, 41, 45, 47, 54, 56, 57, 72, 80, 89, 91, 95, 98, 128
Continuous Delivery Pipeline, 27, 37, 38, 39, 47, 56, 108, 118, 128
contracts, 42, 43, 44, 68
coverage, 17, 18, 19, 20, 22, 23, 25, 28, 30, 41, 60, 95
Coverity, 27, 30, 95
credentials, 80, 103
customer, 18, 58, 59, 72, 126
Dashboard, 24, 47, 71, 86, 88
DAST, 29, 30, 55, 69, 97, 119
DDD, 14, 35, 42
dependencies, 94, 96
deployment, 19, 29, 30, 31, 32, 33, 64, 82, 87, 120, 128
Dev, 15, 19, 20, 23, 28, 30, 31, 33, 47, 52, 53, 54, 56, 57, 64, 67, 71, 80, 89, 103, 104, 108, 110, 120
devices, 31, 32, 33, 34, 35, 84, 95, 118
DevSecOps, 1, 2, 10, 11, 52, 59, 61, 64, 89, 94, 95, 108, 109, 110, 126
DoD, 56, 63, 90, 91
domain, 14, 15, 24, 25, 27, 28, 45, 47, 92, 96, 118
Domain Model, 16, 26, 27, 28, 34, 37, 68, 69, 81, 97, 98, 111
downstream, 19, 20, 41, 43, 85
DSL, 45, 93, 94
E2E, 21, 43, 44, 47, 48, 49, 120
embedded systems, 34
environment, 21, 28, 29, 30, 31, 32, 33, 47, 54, 86, 96, 103, 120
ephemeral, 84, 96, 99, 101, 102, 120, 124
farm, 25, 71, 88, 95, 96

feature, 57, 83, 84
FeatureLeadTime, 62, 63, 110
firmware, 14, 34
Functional, 18, 21, 22, 29, 30, 54, 55, 66, 69, 82, 119
gates, 19, 20, 21, 23, 39, 61, 69, 70, 77, 93, 94, 97, 116, 118, 119, 121
Gradle, 30, 33
headless, 19, 20, 21, 23, 73, 79
hierarchy, 106
HockeyApp, 31, 33
IaaS, 46, 94
immutable, 95, 120
index, 58, 112
InfoSec, 12, 58, 60, 108, 111
infrastructure, 39, 83, 89, 95, 120
insights, 11, 51, 53, 63, 85, 104
Integration, 10, 12, 15, 22, 39, 54, 55, 66, 69, 81, 82, 98, 119, 128
iOS, 25, 27, 31, 32, 41, 95
Java, 25, 27, 29, 41, 95
Jenkins, 2, 45, 94, 124
Jira, 25, 29, 30, 31, 32, 33, 40, 71, 72, 88, 96
JUnit, 30, 32, 76, 95
Key Takeaways, 13, 35, 50, 63, 91, 105, 115, 127
KPIs, 24, 47, 58, 62, 64, 71, 88, 89, 96, 104, 110
latency, 70, 89
metric, 58, 60
microservices, 21, 26, 47, 48, 49
Microsoft, 46
mobile, 25, 35, 71, 84, 88, 95, 96
mocks, 17, 20, 34, 49
model, 14, 15, 24, 26, 27, 28, 37, 38, 45, 86, 89, 101
monolithic, 21, 47
MTTR, 59, 61, 65
MVP, 65, 66, 76
network, 64, 70, 89
Nexus, 28, 30, 47, 79
NodeJS, 25, 27, 28, 41, 95
notifications, 70, 82, 84, 104
Operations, 61, 108, 109, 120
orchestrators, 76, 94

OSS, 84, 90, 93, 94
OWASP, 29, 30, 95
PaaS, 24, 46, 71, 88, 89, 94, 95, 96
pattern, 21, 43, 47, 53, 87
patterns, 11, 49, 51
performance, 17, 19, 20, 25, 29, 30, 39, 43, 44, 47, 54, 55, 67, 69, 71, 72, 81, 82, 83, 88, 93, 95, 96, 97, 104, 119, 120
PII, 74, 116
Pipeline, 2, 14, 15, 16, 17, 18, 19, 20, 21, 22, 23, 24, 25, 26, 27, 28, 29, 31, 32, 34, 35, 36, 37, 39, 41, 42, 43, 44, 45, 46, 47, 54, 55, 56, 57, 59, 61, 63, 65, 66, 68, 69, 70, 71, 72, 73, 74, 75, 76, 77, 78, 79, 80, 81, 82, 83, 84, 87, 88, 89, 90, 91, 92, 93, 94, 95, 96, 97, 98, 99, 100, 101, 102, 103, 104, 105, 108, 110, 111, 112, 113, 114, 115, 118, 119, 120, 121, 123, 124
Pipeline-As-Code, 45, 94, 98, 115, 121
Pipelines, 21, 25, 28, 41, 43, 45, 47, 55, 56, 57, 59, 61, 72, 73, 85, 92, 94, 96, 97, 98, 99, 101, 102, 104, 106, 110, 124
polyglot, 26, 93, 96
predictability, 21, 61, 77, 106, 113, 120
process, 11, 101, 108, 121, 126
Process-As-Code, 39, 92, 99, 105
processes, 10, 11, 24, 99, 104, 122
Producer, 41, 42, 43, 44
Production, 12, 15, 19, 21, 22, 23, 25, 28, 29, 31, 32, 33, 35, 39, 46, 47, 49, 52, 53, 54, 56, 61, 65, 67, 71, 73, 75, 77, 79, 80, 83, 84, 89, 90, 103, 104, 110, 119, 120, 121
productivity, 92, 104, 122
quality, 11, 12, 15, 21, 39, 53, 54, 55, 59, 61, 77, 92, 108, 120
RBA, 93, 94, 97, 116
Release, 11, 98, 108, 109, 128
repository, 12, 15, 20, 21, 22, 23, 24, 26, 28, 30, 31, 32, 45, 46, 47, 71, 79, 80, 82, 88, 93, 96, 102, 121
RoI, 48, 66, 103
rollback, 72, 84, 90
S3, 28, 30, 47, 79
SaaS, 25, 46, 71, 88, 96, 126

SAST, 17, 18, 20, 22, 23, 28, 30, 55, 66, 69, 95, 119
SauceLabs, 25, 29, 31, 32, 33, 41, 71, 88, 95, 96
SCA, 66, 119
scalability, 99, 101
secrets, 82, 103
security, 11, 17, 18, 19, 20, 22, 23, 25, 28, 29, 30, 39, 41, 55, 59, 60, 61, 64, 66, 69, 73, 74, 80, 81, 82, 86, 95, 97, 108, 110, 118, 119
services, 22, 24, 25, 26, 28, 49, 77, 93, 95, 96, 97, 101
SLA, 26, 78, 84, 87
SoD, 106, 107, 108, 110, 111, 112, 114, 115
Sprints, 53, 58, 75
Stage, 15, 19, 21, 23, 25, 29, 30, 32, 33, 47, 53, 54, 56, 61, 67, 71, 77, 80, 81, 89, 104, 110, 120, 123
stateless, 99, 103
static code analysis, 17, 18, 19, 20, 22, 23, 28, 30, 31, 33, 42, 43, 44, 54, 69, 80, 82, 95, 119
Statistics, 69, 70
subsystem, 22, 38, 56, 86
subsystems, 19, 20, 21, 47, 48, 68, 81, 86, 97
SumoLogic, 24, 47, 71, 88, 96, 104
system, 15, 20, 21, 68, 81
systems, 14, 24, 25, 46, 71, 72, 89
test, 28, 30, 54, 56, 60, 66, 69, 72, 76, 77, 78, 82, 90, 97, 120, 126
TestNG, 30, 76, 95
toggles, 72, 74, 75, 83, 84
traceable, 45, 68, 80, 84, 94, 96
Twelve-Factor, 92, 128
ubiquitous language, 38, 45
Unit, 54, 55, 60, 66, 69
upstream, 41, 43
velocity, 26, 39, 49, 57, 58, 65, 66, 92, 104, 111, 122
vendors, 46, 84, 86, 94, 126
versioned, 15, 19, 20, 21, 22, 23, 24, 28, 30, 31, 32, 33, 42, 43, 44, 47, 71, 84, 88, 96, 98, 120
VSM, 34, 121, 122
ZAP, 29, 30, 95
ZDD, 23, 87

Chapter 1 | Introduction

I have had the privilege to witness organizations go through transformations of culture that build trust and improve efficiency. I have had the pleasure of working with some of the brightest minds and have had the good fortune of seeing my design decisions stand the test of time, and in some cases not as much.

Processes, amongst other things, can exponentially improve organizational culture. This book lays the foundation of building resilient processes for transformational leaders who aspire to drive culture change in their organizations. The principles of DevSecOps, Continuous Integration and Continuous Delivery form the recipe for success, and Continuous Improvement is at the heart of it.

As you plow through the chapters of this book, you will virtually experience the thrills and the shivers of leading crucial transformational changes in enterprises. And you will be able to define the next chapter of your very own journey.

So, are you ready to find your true north? Then, fasten your seatbelts!

DevSecOps: The New DevOps?

DevSecOps. Ah wait, wasn't this DevOps before? Did we forget security the first time around?

If we want to stay in business, we would not dare to forget security. This time around, we want to double up and ensure that we build security into our applications from the design phase, instead of evaluating a finished product. We are doing a "*Left Shift*" by pulling security upfront.

Shouldn't this be SecDevOps, then? Or DevQASecOps? Did we drop Release out? And what about Platform, Infrastructure, and Tools?

We have adopted every function that's critical to delivering quality software frequently and predictably. To win, everyone should come together to design quality, security and operational excellence into our processes. Cliché or not, "United we stand, divided we fall".

And, what if our Engineering organization is severely fragmented today? Then we should start investing in DevSecOps principles in a collaborative fashion irrespective of administrative divides. There is no "right" organizational structure. Eventually, we will discover a framework that's right for our organization, whereby decisions flow fast and brilliant vision meets laser-fast execution.

Which brings us to the million-dollar question. Where does process fit in the organizational hierarchy? Process should have a first-class seat at the table, since process drives architecture, and architecture drives tools. If we decide on tools first, we get stuck with what the chosen tools can or cannot do. Moreover, tools and platforms have gone through revolving doors, driving some people to despair. Understand that evolution of tools is inevitable, and the design changes have been for the greater good of humanity. As long as we keep our eyes on the ball, we are good. And, that ball is process.

Also, we should focus on where the ball is going to drop instead of where the ball was thrown. Which means, automating today's processes is great as long as we keep an eye on what the process should evolve to for the long-term sustainability of our business. For example, invest in value stream mapping, and get rid of "waste". Do not automate it, just because you can.

This book shares direct insights, articulates design patterns and touches on technical recipes that have been used in the industry. Wherever we are in our journey of evolution or revolution or renaissance, this book aims to make that journey rewarding and fulfilling for us.

So, buckle up!

Continuous Everything. And Everyone.

Continuous Integration or CI is a core technique where everyone commits every day and code is integrated early and often.

Continuous Improvement and *Continuous Innovation* are referred to as CI too. For whatever it's worth, Continuous Integration, Continuous Improvement and Continuous Innovation are in the same spirit.

Continuous Delivery or CD stems from Continuous Integration (CI), where as per business needs, quality products are released frequently and predictably from source control repository to Production in an automated fashion. This improves developer feedback and reduces shelf-time of new ideas, and thereby improves the sustainability of businesses.

Continuous Deployment, also known as CD, has the goodness of Continuous Delivery, and additionally enables us to deploy to Production without the need for a business approval or a manual gate. This skyrockets Time2Market and enables rapid experimentation to outmaneuver competition.

Last but not the least, *Continuous Testing* is a core subset of Continuous Integration, Delivery and Deployment and is the secret sauce without which the entire Continuous paradigm would crumble to pieces.

Sooner or later, everyone in the organization gets a call when Continuous Delivery Pipelines are being constructed. Executives, Engineering, Product, Governance, Risk, Compliance, InfoSec, Operations, Legal and whatever you have. All of us are part of this transformation, one way or the other, and *Continuous Everyone* is a thing now!

Typically, we humans tend to procrastinate what's difficult. And software releases are occasionally seen as painful. However, the only way to cut ahead is to do it over and over again, till it's trivial. And boring. Throughout this book, we will invest into designing continuity into our processes, so that we can reduce difficulty by increasing frequency of our product releases, till the releases become uneventful.

Acknowledgement

Thank you, Martin Fowler, Jez Humble and Eric Evans, for the awesome resources you provide to enable self-learning.

I am grateful to my friends, colleagues and partners at LifeLock, Gap Inc., Yahoo!, Apple, GoPro, ThoughtWorks, CloudBees, Sonatype, ElectricCloud, TechBeacon, PricewaterhouseCoopers Ltd. and Walmart for the wonderful opportunities that I shared with you folks in "Doing The Right Thing".

I am thankful to my friends and family for your love and support. Yes, I know. No one was expecting me to write books. In case you are wondering, I wasn't expecting me to write books either.

I am grateful to you, Dipanjan Ghosh, for you are a repeat offender in helping me pull this second book through. As if the first wasn't enough of a carnival, I continue to celebrate going through the author's journey with you.

And, last but not the least, a sincere note of thanks to my readers for taking time out of your busy schedules and accompanying me on this journey. The community is very special to me, and I hope you will stay engaged as we untie the knots together.

KEY TAKEAWAYS FROM CHAPTER 1

- When is a good time to start Continuous Delivery? It was yesterday. Run, reader, run. It's fun to run!

- Set aside some cash and get a priority from your C-Suite, so that execution is uninterrupted.

- You can't figure everything out on Day 1. No one can. And that's okay. Write down your assumptions, and learn and iterate as you go.

CHAPTER 2 | CONTINUOUS DELIVERY AS A DOMAIN

Domain-driven design (DDD) is an approach to software development for complex needs that connects the implementation to an evolving model. Domain-driven design (DDD) was authored by Eric Evans and has widespread adoption in the industry.

DDD is known to be overkill for simple problems. After careful consideration of the issues that plague product releases (software, firmware, embedded systems, Internet of Things, *aaS, infrastructure, metrics, and the like), I decided to model the Continuous Delivery Pipeline as a domain.

I will be touching on DDD principles to explain how they apply to Continuous Delivery; however, my readers are encouraged to refer to the fantastic literature authored by Eric Evans.

Domain

The domain is a sphere of knowledge, influence, or activity. The subject area to which the user applies a program is the domain of the software.

In this chapter, we will treat Continuous Delivery as our domain.

Continuous Delivery is a core technique:

- That stems from Continuous Integration

- Where as per business needs, quality products are released frequently and predictably from source control repository to Production in an automated fashion through a Pipeline.

Domain Events

Information about activity in the domain is modeled as a series of discrete events. A domain event is a full-fledged part of the domain model, a representation of something that happened in the domain.

Domain events:

- Are events that domain experts care for

- Are distinct from system events that reflect activity within the software itself. System events could be associated with domain events.

A few examples of domain events in the Continuous Delivery domain are:

- Publication of versioned artifacts to the artifact repository by the Pipeline

- Promotion of versioned artifacts by the Pipeline for downstream consumption

    - From Dev to Stage

    - From Stage to Production

- Submission of Audit Trail (or evidence) by the Pipeline to a Change Management System or System of Record

Domain Model For Continuous Delivery Pipeline

Let's break down the larger problem into multiple smaller problems. The Pipeline has several phases that have different objectives. A common misconception is to perceive these phases as physical. They are logical phases that enable seamless delivery of products through software gates that either allow or disallow promotion of versioned artifacts from one phase to the next.

In this chapter, we will be touching on six phases. It is up to you on how you would like to incorporate these phases in your own organization. There could be more than six phases, or fewer. The overarching goal here is to understand the objective of each phase, and to design Pipelines that have segregation of duties built in.

Phase 1: Pre-Develop Commit Phase

Objective

This phase validates code that resides in a Feature Branch before it is merged into the team's Develop Branch.

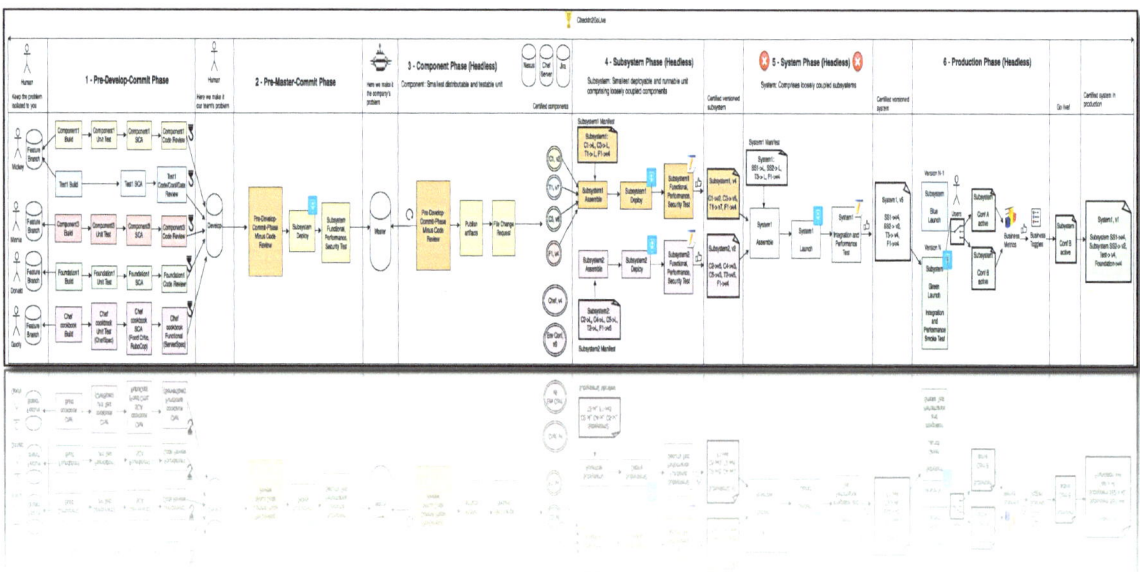

Figure 1 - Domain Model For Continuous Delivery Pipeline

View, print and download from http://continuity.world

The Feature Branch is an individual developer's branch and the Develop Branch is a team branch where developers in a Scrum Team merge code. It is better to solve individual problems in isolation in the Feature Branch before making it the team's problem, since when something breaks in the Develop Branch, the entire team could get blocked from going to Master, and that's expensive.

Caveat

As far as branches go, less the merrier.

Branching is an overhead, and hence we should carefully assess the branching strategy to fit our specific needs. We can combine the Pre-Develop-Commit Phase with Pre-Master-Commit Phase into a single Pre-Commit Phase, thus eliminating the need for both Feature AND Develop branches. Or, we could check into Master directly, if we can nurture that discipline into our teams. The overarching goal here is that Master should not break too often and we need to ensure that in the most effective way with the least overhead of branches.

Lowest Common Denominator For Exit Criteria

Minimum requirements for the Pipeline to promote code from this phase to the next are:

- Build

- Execution of unit tests and measurement of code coverage

- Static code analysis with linters to detect deviations from coding best practices

- Static code analysis to detect security vulnerabilities or Static Analysis Security Testing (SAST)

On top of this, it is a good practice to run functional, security and performance tests in containers and with mocks.

Phase 2: Pre-Master Commit Phase

Objective

This phase validates code that resides in the Scrum Team's Develop Branch before merging it into the Master Branch.

The Develop Branch is a team branch where developers in a Scrum Team check in code, whereas the Master Branch is where the organization's code comes together. It is better to solve team problems in isolation in the Develop Branch before making it the organization's problem. When the Develop Branch breaks, that particular team is inconvenienced, however, when the Master branch breaks, the customer could potentially stop receiving feature updates and bug fixes, and that's fatal.

Caveat

As far as branches go, less the merrier.

Branching is an overhead, and hence we should carefully assess the branching strategy to fit our specific needs. We can combine the Pre-Develop-Commit Phase with Pre-Master-Commit Phase into a single Pre-Commit Phase, thus eliminating the need for both Feature AND Develop branches. Or, we could check into Master directly, if we can nurture that discipline into our teams. The overarching goal here is that Master should not break too often and we need to ensure that in the most effective way and the least overhead of branches.

Lowest Common Denominator For Exit Criteria

Minimum requirements for the Pipeline to promote code from this phase to the next are:

- Build

- Execution of unit tests and measurement of code coverage

- Static code analysis with linters to detect deviations from coding best practices

- Static code analysis to detect security vulnerabilities or Static Analysis Security Testing (SAST)

- Deployment

- Parallel execution of:

    o Functional Tests

- Performance Tests

- Dynamic Analysis Security Testing (DAST)

The overarching goal of this phase is to ensure that Master doesn't break too often.

Phase 3: Component Phase

Objective

This is where components (the smallest distributable and testable units) are built and tested before distribution to downstream subsystems (the smallest deployable and runnable units made from loosely coupled components). The Pipeline runs the show in a headless fashion with automated agents and software gates.

Caveat

This is a logical phase and not a physical one, as often misunderstood by organizations. This is a precursor for Subsystem Phase and is often combined with Subsystem Phase in the same environment that most of the industry refers to as "Dev environment" or "QA environment".

We may choose to defer publication of versioned artifacts till after the Subsystem Phase is complete, since published artifacts will be of higher confidence once they have been deployed and tested for functionality, performance and security. However, some organizations prefer to download artifacts from the artifact repository for deployment and testing, since that's how it will be done in Stage and Production, in which case publication of artifacts becomes part of this Component Phase.

Lowest Common Denominator For Exit Criteria

Minimum requirements for the Pipeline to promote code from this phase to the next are:

- Build

- Execution of unit tests and measurement of code coverage

- Static code analysis with linters to detect deviations from coding best practices

- Static code analysis to detect security vulnerabilities or SAST (Static Analysis Security Testing)

- Publication of versioned artifacts to the artifact repository

On top of this, it is a good practice to run functional, security and performance tests in containers and with mocks.

Phase 4: Subsystem Phase

Objective

This is where subsystems (the smallest deployable and runnable units made from loosely coupled components) are built and tested before distribution to the downstream system (loosely coupled subsystems). The Pipeline runs the show in a headless fashion with automated agents and software gates.

Caveat

This is a logical phase and not a physical one, as often misunderstood by organizations. This is a precursor for System Phase and is often combined with Component Phase in the same environment that most of the industry refers to as "Dev environment" or "QA environment".

Lowest Common Denominator For Exit Criteria

Minimum requirements for the Pipeline to promote code from this phase to the next are:

- Build

- Execution of unit tests and measurement of code coverage

- Static code analysis with linters to detect deviations from coding best practices

- Static code analysis to detect security vulnerabilities or Static Analysis Security Testing (SAST)

- Publication of versioned artifacts to the artifact repository

- Deployment of downloaded versioned artifacts from the artifact repository

- Parallel execution of:

    o Functional Tests

    o Performance Tests

    o Dynamic Analysis Security Tests (DAST)

Phase 5: System Phase

Objective

This is where a system is assembled from loosely coupled subsystems (the smallest deployable and runnable units) and validated before the system assembly is promoted to Production. The Pipeline runs the show in a headless fashion with automated agents and software gates.

Caveat

This System Phase demonstrates the composition anti-pattern by assembling loosely coupled subsystems into one system that can be launched only as a whole. This in turn results in a highly integrated environment, where subsystems are tied at their hips for success. This all-or-none software delivery approach is less than admirable, however, the Pipeline is a mere reflection of the highly coupled architecture of the product that's flowing through the Pipeline. Note that the Pipeline can surface architecture coupling issues with rapid precision, but cannot solve them for you.

Enterprises tend to have a highly coupled monolithic legacy system architecture that requires composition (or coupling) of subsystems. Intense integration requirements lead to questionable speed, quality and predictability, and this phase has been the nemesis of most Continuous Delivery Pipelines in enterprises. This phase correlates well to what the industry refers to as the "E2E (End to End)" environment.

In microservices architecture, loosely coupled subsystems do not need to be assembled into a monolithic system. Subsystems could arrive on Stage and go along their merry way to Production, as long as they can work with old and new versions of neighboring subsystems. Microservices are independently deployable artifacts, and hence are subsystems by definition, that is, they are the smallest deployable and runnable units.

However, microservices architecture needs just as good integration tests due to the much higher granularity of services. Additionally, integration tests need to account for both forward and backward compatibility, since it makes business sense to let the faster subsystems run ahead of the slower ones.

Lowest Common Denominator For Exit Criteria

Minimum requirements for the Pipeline to promote code from this phase to the next are:

- Build

- Execution of unit tests and measurement of code coverage

- Static code analysis with linters to detect deviations from coding best practices

- Static code analysis to detect security vulnerabilities or Static Analysis Security Testing (SAST)

- Publication of versioned artifacts to the artifact repository

- Deployment of downloaded versioned artifacts from the artifact repository

- Parallel execution of:

    - Functional Tests

    - Integration Tests

    - Performance Tests

    - Dynamic Analysis Security Testing (DAST)

Phase 6: Production Phase

Objective

Upon successful execution of relevant tests, the subsystem (or system) gets promoted from the Subsystem (or System) phase to Production phase with no further modifications. The Pipeline runs the show in a headless fashion with automated agents and software gates.

Caveat

Although the Lowest Common Denominator for earlier phases does not explicitly state it, ZDD (Zero Downtime Deployment) could and should be practiced in Dev and Stage, just as much as in Production. A/B tests should be executed earlier too, to get higher confidence in the product under development.

The difference between "*deployment*" and "*launch*" is interesting. "Deployment" involves moving the bits and "launch" involves turning them on, either for the whole population or selectively for a cross-section. "*Dark Launches*" is a term that has been made popular by companies who discreetly evaluate new features without drawing the public or the media's attention. They like to experiment with new features in a real Production environment on a small cross-section of the population. If things go south, the bits are turned off, and the disruption is minimal. If business KPIs are met, they roll the new features out to the masses at small pulse increments of 5-10%, all the way to 100%.

Lowest Common Denominator For Exit Criteria

Minimum requirements for the Pipeline to turn the product live in Production are:

- Build

- Execution of unit tests and measurement of code coverage

- Static code analysis with linters to detect deviations from coding best practices

- Static code analysis to detect security vulnerabilities or Static Analysis Security Testing (SAST)

- Publication of versioned artifacts to the artifact repository

- ZDD (Zero Downtime Deployment) of downloaded versioned artifacts from the artifact repository. One approach could be Blue-Green deployments.

- Parallel execution of Integration, Performance and Security smoke tests

- A/B Tests to evaluate business KPIs (Key Performance Indicators)

- Product Launch

Domain Services

These services are significant processes in the domain, which are not natural responsibilities of an entity or value object.

We can add operations to the domain model as a standalone interface declared as a service.

Standard use cases where the Pipeline interfaces with external systems are:

- A source control repository like Atlassian's Bitbucket or GitHub

- A dashboard to log KPIs (Key Performance Indicators) that enable the leadership team to make informed decisions, like SumoLogic or ELK Stack

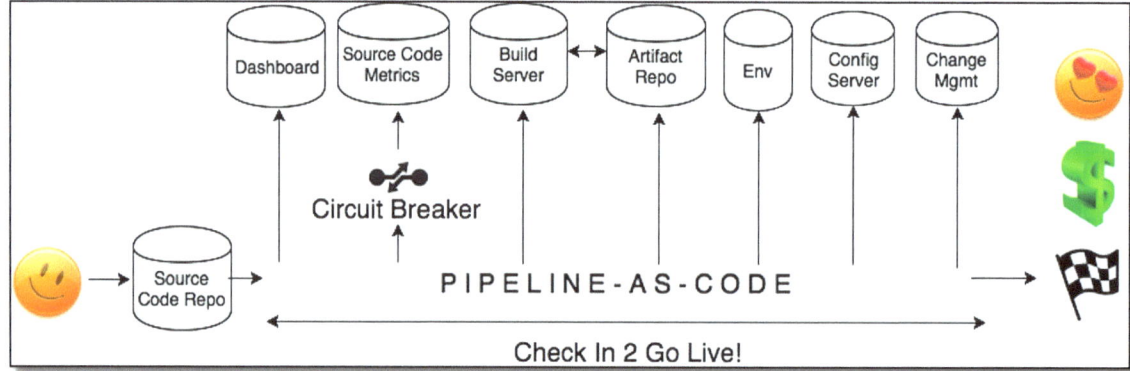

Figure 2 - Domain Services For Continuous Delivery Pipeline

View, print and download from http://continuity.world

- An artifact repository to publish versioned artifacts, like Sonatype's Nexus Repository or JFrog's Artifactory or Amazon's S3

- On-prem data center VMWare VMs or Amazon's AWS or Microsoft Azure or Pivotal's Cloud Foundry or Heroku or a similar Platform as a service (PaaS) to deploy and execute tests

- Atlassian's Jira to manage bugs and change requests

- A configuration server to manage configuration

- A web-browser and mobile device farm to execute UI functional tests, like SauceLabs or Amazon Device Farm

- An external SaaS service like BlazeMeter to run performance tests

- Etc.

These services are shared assets between different Pipelines. The interfaces with these external systems should be sharp and wrapped into services.

Based off the above interfaces of the Pipeline with external systems, domain services could be defined around:

- CreateChangeRequest (), TransitionChangeRequest (), CloseChangeRequest () etc. to manage the change request lifecycle in Jira

  Creating audit trails is a compliance requirement, and unless AWS Cloud Trail suffices, every Pipeline will need to call into these services, and manage change requests for deployments to Production and, in some cases, Stage too.

- EstablishSecureTunnel () to execute functional tests and gather results through a tunnel that will be established between the Pipeline and a SauceLabs VM in the SauceLabs Data Center

  NodeJS, iOS and Android Pipelines will call into this service to execute functional tests and gather results without jeopardizing security.

- GenerateCodeCoverageReport () to write and gather code coverage metrics using Cobertura or a similar tool, that can be graphed/trended

  NodeJS, Java and other Pipelines will call into this service.

The Services Architecture comes in handy to define domain services for the Pipeline.

Pipeline domain services:

- Could be made of a suite of evolving services that can potentially be developed by anyone in the organization, instead of by one dedicated team of service providers.

- Allow freedom to Scrum Teams to opt in and out of services that offer little benefit, other than mandatory compliance requirements. In general, Scrum

Teams should be allowed to control their own destiny so that they are accountable for their own velocity and productivity.

- Should define sharp service interfaces that declare the exact intent of the service. Each service should be granular and should address one single business capability, much like how microservices are defined to do one small thing well.

- Should maintain existing interfaces, such that Scrum Teams experience uninterrupted service. Upgrades should not cause disruption and wreak havoc with Pipeline SLAs (Service Level Agreements). See section on "Anticorruption Layer" in chapter "Pipeline Domain Model Integrity".

- Should not lock down Scrum Teams to use future service offerings based off the usage of current services. Also, we should avoid vendor lock-ins.

- Should not force Scrum Teams to upgrade to a newer version of the service unless they would like to avail newer features.

- Pipeline services reside in a shared repository, which allows commits based on pull requests and code reviews. The Pipeline for the Pipeline needs to validate new releases of domain services before they are made available for consumption. See section on "Continuous Integration" in chapter "Pipeline Domain Model Integrity".

- New services should be registered to enable service discovery and prevent duplication.

- New services should be discoverable.

Depending on the exact nature of these services, domain services can be implemented as web services, REST services, AWS Lambda functions stitched together with AWS Step Functions, static polyglot functions using the Services Architecture or whatever makes sense for your organization.

Pipeline Domain Model Fragmentation

In an ideal world, we would have a single, unified model. This is aspirational at best and is more an academic goal than one that makes sense in the industry.

A more practical perspective would be to allow fragmentation of the model into multiple models, wherever it makes sense. Product owners and architects walk the fine line regarding how many fragments is too many. Fragmentation makes execution easier for teams who strive to connect the implementation to the evolving model.

Figure 3 demonstrates the domain model for a Continuous Delivery Pipeline that is fragmented for different languages and platforms, like iOS, Android, NodeJS, Java, and SQL etc. We could have more fragments based on the tech stack that we standardize on in our organization. The leaner our tech stack, the easier to build and maintain Pipelines.

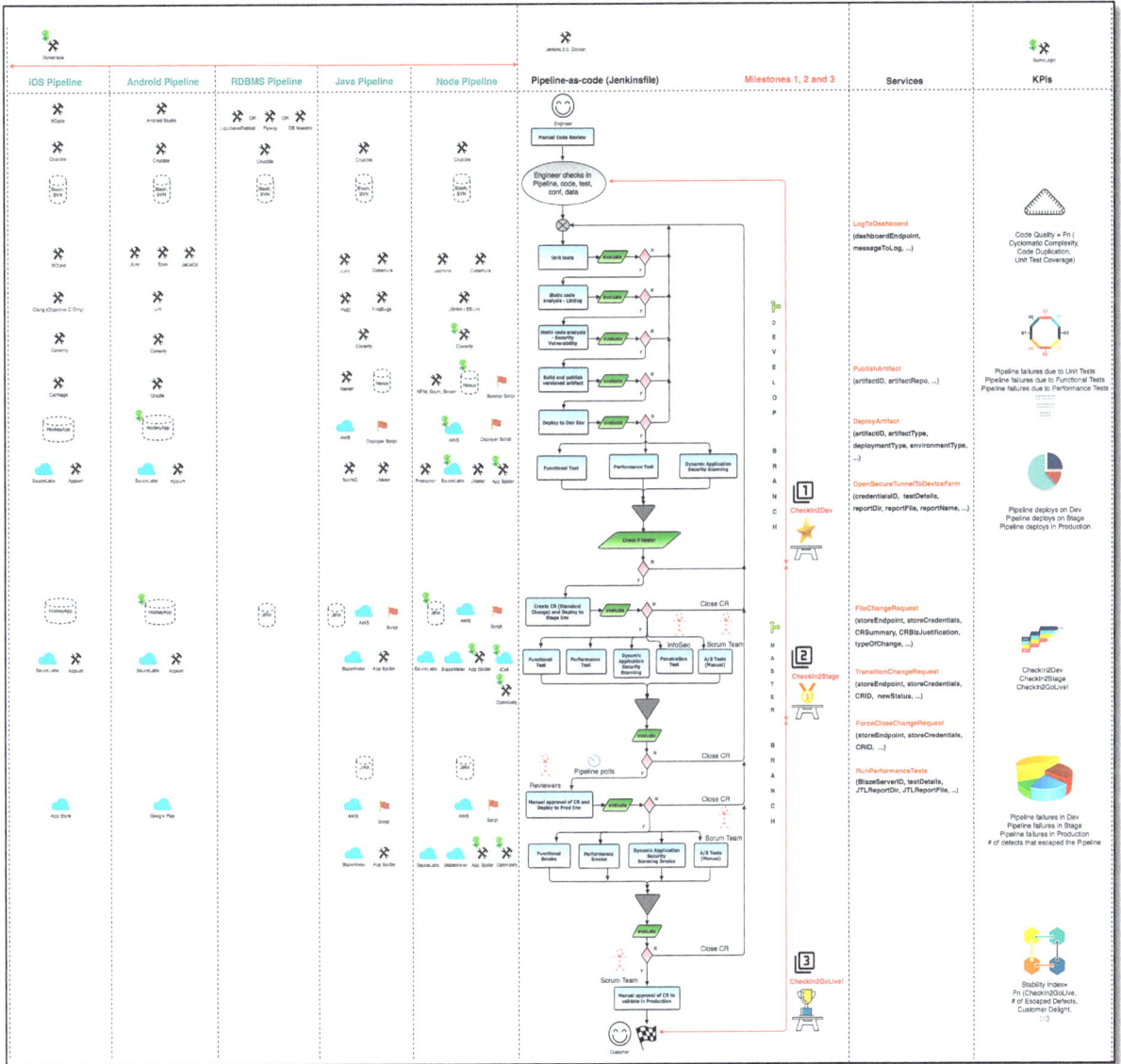

Figure 3 - Fragmentation Of Pipeline Domain Model

View, print and download from http://continuity.world

The product architect infuses the principles for:

- Maintaining model integrity. See chapter "Pipeline Domain Model Integrity" for details.

- Working with multiple models at the same time, while honoring shared domain services.

Several teams could be connecting their Pipeline implementations to this fragmented model at the same time. They should be re-using domain services, which are shared between the various fragments of the Pipeline domain model.

The following sections walk through fragmentations based on languages or platforms that are used in organizations of different sizes and shapes to build Continuous Delivery Pipelines. These pipelines can deliver software from source code repository to Production with one manual gate right before deployment to Production.

NodeJS Pipeline Model

A NodeJS Pipeline can be built to:

- Access code from source code repository like Atlassian's Bitbucket or GitHub

- Have code review integrations with tools like Crucible

- Run unit tests using Jasmine

- Measure unit test coverage using Cobertura

- Run static code analysis using JSLint / ESHint

- Run SAST (Static Analysis Security Testing) with Coverity

- Build versioned artifacts using NPM, Grunt, Bower

- Publish versioned artifacts to Sonatype's Nexus Repository or JFrog's Artifactory or Amazon's S3

- Deploy to a Dev environment like Amazon EC2

- Run tests on Dev in parallel:

- o   Functional tests using Protractor on SauceLabs

- o   Performance tests using BlazeMeter

- o   Security scans using OWASP (Open Web Application Security Project) ZAP

- Create a standard change request in Jira to signify an impending deployment to Stage

- Deploy to a Stage environment like Amazon EC2

- Run tests on Stage in parallel:

    - o   Functional tests using Protractor on SauceLabs

    - o   Performance tests using BlazeMeter

    - o   Dynamic Analysis Security Testing (DAST) using OWASP (Open Web Application Security Project) ZAP

    - o   Penetration tests using tCell

- Advance the standard change request in Jira to signify an impending deployment to Production

- Deploy to a Production environment like Amazon EC2

- Run smoke tests in parallel in Production

- Advance the standard change request in Jira to request human validation in Production

- Have a human close the change request after validating manually in Production

Java Pipeline Model

A Java Pipeline can be built to:

- Access code from source code repository like Atlassian's Bitbucket or GitHub

- Have code review integrations with tools like Crucible

- Run unit tests using JUnit

- Measure unit test coverage using Cobertura

- Run static code analysis using PMD or FindBugs

- Run SAST (Static Analysis Security Testing) with Coverity

- Build versioned artifacts using Maven or Gradle

- Publish versioned artifacts to Sonatype's Nexus Repository or JFrog's Artifactory or Amazon's S3

- Deploy to a Dev environment like Amazon EC2

- Run tests on Dev in parallel:

    o Functional tests using TestNG

    o Performance tests for APIs using BlazeMeter

    o Security scans using OWASP (Open Web Application Security Project) ZAP

- Create a standard change request in Jira to signify an impending deployment to Stage

- Deploy to a Stage environment like Amazon EC2

- Run tests on Stage in parallel:

    o Functional tests using TestNG

    o Performance tests for Java APIs using BlazeMeter

    o DAST (Dynamic Analysis Security Testing) using OWASP (Open Web Application Security Project) ZAP

- o   Penetration tests using tCell

- Advance the standard change request in Jira to signify an impending deployment to Production

- Deploy to a Production environment like Amazon EC2

- Run smoke tests in parallel in Production

- Advance the standard change request in Jira to request human validation in Production

- Have a human close the change request after validating manually in Production

iOS Pipeline Model

An iOS Pipeline can be built to:

- Be aware of the XCode development platform

- Access code from source code repository like Atlassian's Bitbucket or GitHub

- Have code review integrations with tools like Crucible

- Run unit tests using XCUnit

- Run static code analysis using Clang (Objective C) and Tailor (Swift)

- Build versioned artifacts signed with Dev certificates using Carthage

- Publish versioned mobile apps (*.ipa) to HockeyApp

- Deploy to a Dev environment on device farms like Amazon Device Farm or SauceLabs or local mobile devices

- Run functional tests using Appium on device farms like Amazon Device Farm or SauceLabs or local mobile devices

- Create a standard change request in Jira to signify an impending deployment to Stage

- Deploy a versioned app signed with the Stage certificate to a Stage environment on device farms like Amazon Device Farm or SauceLabs or local devices

- Run functional tests using Appium on device farms like Amazon Device Farm or SauceLabs or local mobile devices

- Advance the standard change request in Jira to signify an impending deployment to Production

- Deploy a versioned app signed with the Production certificate to a Production environment on device farms like Amazon Device Farm or SauceLabs or local devices

- Run smoke tests in Production

- Upload app to Apple's App Store

- Advance the standard change request in Jira to request human validation in Production

- Publish app (could be manual) in App Store and make it available for iOS customers

- Have a human close the change request after validating manually in Production

Android Pipeline Model

An Android Pipeline can be built to:

- Be aware of the Android Studio development platform

- Access code from source code repository like Atlassian's Bitbucket or GitHub

- Have code review integrations with tools like Crucible

- Run unit tests using JUnit, Spek etc.

- Measure unit test coverage with JaCoCo

- Run static code analysis using Lint

- Build versioned artifacts using Gradle signed with Dev certificates

- Publish versioned apps (*.apk) to HockeyApp

- Deploy to a Dev environment on device farms like Amazon Device Farm or SauceLabs or local devices

- Run functional tests using Appium on device farms like Amazon Device Farm or SauceLabs or local mobile devices

- Create a standard change request in Jira to signify an impending deployment to Stage

- Deploy a versioned app signed with the Stage certificate to a Stage environment on device farms like Amazon Device Farm or SauceLabs or local devices

- Run functional tests using Appium on device farms like Amazon Device Farm or SauceLabs or local mobile devices

- Advance the standard change request in Jira to signify an impending deployment to Production

- Deploy a versioned app signed with the Production certificate to a Production environment on device farms like Amazon Device Farm or SauceLabs or local devices

- Run smoke tests in Production

- Upload app to Google's Play Store

- Advance the standard change request in Jira to request human validation in Production

- Publish app (could be manual) in Play Store and make it available for Android customers

- Have a human close the change request after validating manually in Production

Firmware and Embedded Pipeline Model

A common misconception is that Continuous Delivery applies only to software.

We studied the essence of "Domain Model For Continuous Delivery Pipeline" and the same principles apply just as much to firmware, embedded systems and IoT (Internet of Things). Industries that produce cameras, phones, remote controls, washing machines, traffic signals and the like can benefit from Continuous Integration and Continuous Delivery with some adjustments.

Here are some kinks that we need to be aware of:

Our supply chain is setup in a way that our suppliers refresh devices much later from the time when we request enhancements or fixes. There's a lag that's equivalent to non-value adding waste if we were to do a VSM (Value Stream Map). This has led to some organizations believing that Continuous Delivery is irrelevant wherever hardware is involved.

Figure 4 - Firmware And Embedded Systems

View, print and download from http://continuity.world

See Figure 4. We can use emulators and simulators in the Subsystem Phase and defer using real devices until the System Phase to enable rapid prototyping. This isn't just because it's convenient and less expensive. The rate of change in Component and Subsystem Phases is many folds higher (~50X) than the rate of change in System Phase. Hence emulators and simulators can address the need for speed better. This is equivalent to the concept of using mocks and stubs in software products in the Subsystem Phase and real subsystems in the System Phase.

Real devices should be used in the System Phase, since we need an environment that is identical to the Production environment. The delay in receiving updated versions of real devices could still be high, however, the rate of change would have significantly lessened by that time.

Another way to mitigate this issue is to use 3D printed devices instead of real devices while the churn is high. 3D printers can add to the cost, however, cost-benefit analysis reveals that rapid updates of real devices could be even more expensive.

IoT Pipeline Model

IoT (Internet of Things) could bring on the next Industrial Revolution.

In Figure 5, the egg tray can tell that four out of the initial dozen eggs are remaining, and it could send a fulfillment request to the grocery store ahead of time. It could alert our mobile device too, from which the payment can be processed. Within the next hour or so, a drone could deliver a carton of eggs to our doorstep.

I am humoring an egg-lover like myself, but in general Continuous Delivery concepts apply to most anything we need with adjustments.

Figure 5 – IoT Continuous Delivery of Eggs

View, print and download from http://continuity.world

KEY TAKEAWAYS FROM CHAPTER 2

- DDD (Domain-driven design) principles can be applied to Continuous Delivery Pipelines for software, firmware, embedded systems, IoT and the like.

- Implementation of Continuous Delivery Pipelines can be connected to an evolving and fragmented domain model.

- Domain services form the backbone of the Continuous Delivery Pipeline, and can be designed using the Services Architecture.

Chapter 3 | Pipeline Domain Model Integrity

In an academic setting of model driven development, we try to maintain a consistent and unified model throughout the teams. In reality, there could be many teams who work on different parts of the Continuous Delivery Pipeline ecosystem and the model tends to fragment. However, aspects of the model that are important should stay unified and other parts could be offered for customization.

The product architect walks a fine line on how much fragmentation is too much.

In this chapter, we will discuss ways to maintain the integrity of the Continuous Delivery Pipeline model even while it is fragmenting and evolving.

Bounded Context

The Continuous Delivery Pipeline model, like other models in software development, should be accompanied by the relevant context to which it applies. The context refers to organization, product, code, platform and the like.

External influences should not tamper with the consistency of the model. The devil is in the details though. Product architects find it tricky to define exactly how much "context" is needed to definitively bound a model, and then educate teams so that they can lean in.

Even then, often times there are overlapping contexts and these overlaps can go undetected for a while. This is more likely to happen in larger organizations where no one person is familiar with the entire system.

Ubiquitous Language

We should name each bounded context by a self-explanatory name, and make the names part of ubiquitous language.

We need to define a language that helps the organization communicate effectively. Even though it causes discomfort at first, teams should strive to use this shared terminology within the bounded context. Terminology, when used out of context, could lead to severe misunderstandings and ineffective meetings. For the business, execution delays would translate into unnecessary expenses.

For example:

- A component is the smallest distributable and testable unit
- A subsystem is the smallest deployable and runnable unit
- Etc.

However, do refrain from beating the horse to death by endlessly debating the differences between:

- A component versus a module, or
- A subsystem versus a tier, or

- A system versus a product, or

- A performance test versus a load test versus a stress test, or

- A data quality test versus a data validation test versus a data integrity test

It's more important to agree on any one language and focus on delivering the PoC (proof of concept), than to die a "death by a thousand definitions".

Continuous Integration

Bounded Context keeps the model unified by Continuous Integration.

Continuous Integration and Delivery for the Continuous Integration and Delivery Pipeline are business critical. The same gold standards of product development apply to the Continuous Delivery Pipeline and "Integrate early. Integrate often." is the way to go.

Sounds recursive? Sounds like overkill? Think again. It's not.

The Pipeline is Process-As-Code.

The Pipeline is a first-class citizen, and is an evergreen example of Infrastructure-As-Code.

The Pipeline is a product, which needs checks and balances, much like the products that flow through the Pipeline. Tests for infrastructure get deprioritized in favor of tests for product features and the primary reason is that traditional Product Owners worry less over infrastructure than they do over revenue. However, they care for velocity and productivity, and can be influenced to reserve 20% of the Sprints for infrastructural enhancements and tech debt.

Moreover, checks and balances of a Continuous Delivery Pipeline include tests for the product that's flowing through the Pipeline. If we forget to enable integration or performance or security tests for the product flowing through the Pipeline, we will shoot low quality and insecure code to Production with greater speed till detected.

Also, if the number of tests executing is zero for some reason, we can have the Pipeline abort till an investigation is complete. These concepts tally up into the making of smart software gates, which we will study in details in chapter "Segregation/Separation Of

Duties". These gates are automated, and open or close based off cross-functional formulae agreed upon by our stakeholders.

Context Map

An individual bounded context may suffice for Pipelines in smaller organizations. In enterprises, an individual bounded context leaves some problems in the absence of a global view. Since multiple bounded contexts can exist, it is important to find the interface of each context with one another.

When connections must be made between different contexts, they tend to bleed into each other, unless we are mindful. People on other teams won't be aware of the context bounds and will unknowingly make changes that blur the edges or complicate the interconnections.

One can map the existing terrain that shows the relationship between each bounded context. This provides the much-needed big picture that individual Scrum Teams often lack. That's why the Context Map is a useful tool in Scrum of Scrums. Additionally, once the individual Scrum Teams know where they fit in and how their contributions drive business forward, their engagement is likely to go up.

Shared Kernel

Albeit we should draw clear lines between different contexts, it is sometimes hard to define a boundary. Forcing a boundary for boundary's sake will hurt execution, and the "forced" boundary will stick out like a sore thumb.

It is prudent to identify the overlap of the allied contexts as a shared kernel. A shared kernel should undergo rigorous continuous integration to maintain its coherence between the different contexts. Enhancements will be done infrequently since it has to be agreed upon by members of multiple teams.

A shared kernel can be justified over a context map if participating models are closely related. For example, shared kernels make sense for the following:

- CreateChangeRequest (), TransitionChangeRequest (), CloseChangeRequest () etc. to manage the change request lifecycle in Jira

    All Pipelines are required to create audit trails.

- EstablishSecureTunnel () to execute functional tests and gather results through a secure tunnel that will be established between our Pipeline and a SauceLabs VM in the SauceLabs Data Center

    NodeJS, iOS and Android Pipelines will call into this service to execute functional tests and gather results without jeopardizing security.

- GenerateCodeCoverageReport () to write and gather code coverage metrics using Cobertura or a similar tool, that can be graphed/trended

    NodeJS, Java and other Pipelines will call into this service.

The Pipeline and core pieces of the Pipeline operate like a shared kernel to the rest of the organization. If done wrong, shared kernels could lead to widespread failures and could stall the organization. Hence shared kernels, when needed, should be small and tight.

Customer/Supplier

Context Map relates allied contexts as Customer/Supplier Teams, also known as Producer/Consumer Teams.

In the Continuous Delivery Pipeline domain, there are several instances of downstream priorities factoring into upstream priorities.

Multiple teams get tied at their hips for success and the following complications are frequently observed:

- A Consumer team waits for a Producer team A's work to be completed first.

- Producer team A is unable to complete their work in a timely fashion since the contractual requirements with the Consumer team were unclear. Also, who should drive the contract – the Producer A or the Consumer? And why?

- This Consumer team depends on another Producer team B, who was blissfully unaware of this dependency till late in the game.

- There is no single person who has the whole Context Map flushed out between all Producers and Consumers in the organization.

- Absence of an effective Scrum of Scrums.

This DDD (Domain-driven design) "Customer/Supplier" pattern provides guidelines on how Producers and Consumers should co-exist. In the Intranet World, the Producer is aware of its consumers and needs to honor contracts written by Consumers before it can go live.

The Producer Pipeline should:

- Build the code

- Pass unit tests

- Pass Static Code Analysis checks

- Publish a versioned artifact of the Producer and its double

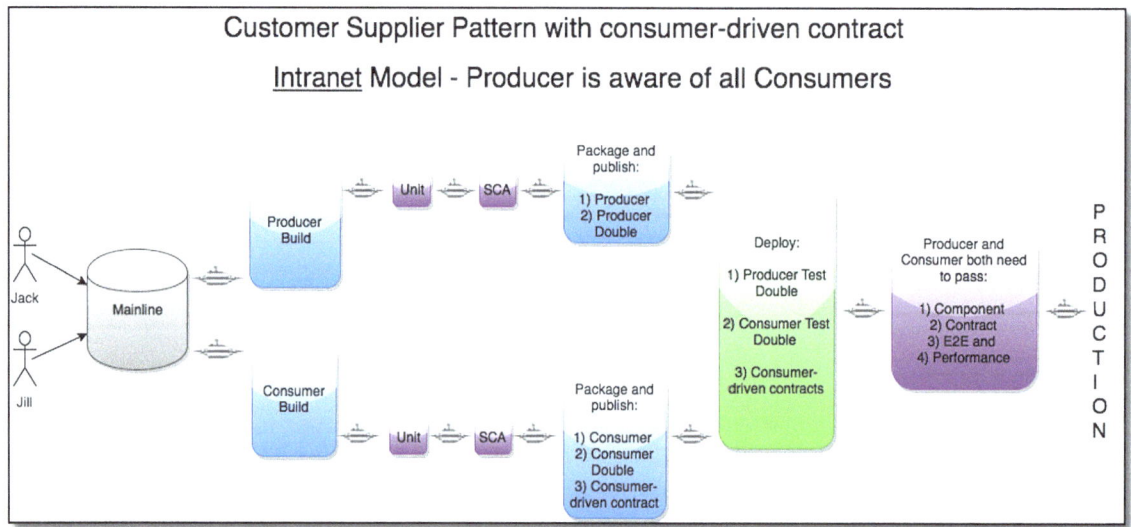

Figure 6 - Customer/Supplier

View, print and download from http://continuity.world

The Consumer Pipeline should:

- Build the code

- Pass unit tests

- Pass Static Code Analysis checks

- Publish a versioned artifact of the Consumer, its double and the Consumer-driven contracts

Both Producer and Consumer Pipelines should then join (or fan-in) to:

- Deploy the Producer's double, the Consumer's double and the Consumer-driven contracts

- Pass component, contract, E2E integration and performance tests

- Go live!

Customer/Supplier teams experience delays in execution if contracts cannot be agreed upon.

Also, note that the Producer owns packaging and distribution of its double, and so does the Consumer. It is dangerous for the Consumer to manufacture a Producer's double, since the Consumer may be unfamiliar with the Producer's latest enhancements, and vice versa. This could lead to false positives or false negatives, thereby leading teams to question the integrity of the Pipeline.

Conformist

In this pattern, there is a unilateral overlap and downstream Consumer jobs are conformists to upstream Producer jobs. The upstream job has no motivation to provide for the downstream job's needs.

In the Internet World, the Producer is unaware of its consumers, and hence cares less about their exact needs. In this pattern, the Producer drives contracts and the Consumer conforms to those contracts.

The Producer Pipeline should:

- Build the code

- Pass unit tests

- Pass Static Code Analysis checks

- Publish a versioned artifact of the Producer, its double and the Producer-driven contracts

- Deploy the Producer's double along with the Consumer's double

- Pass component-level and performance tests

- Go live!

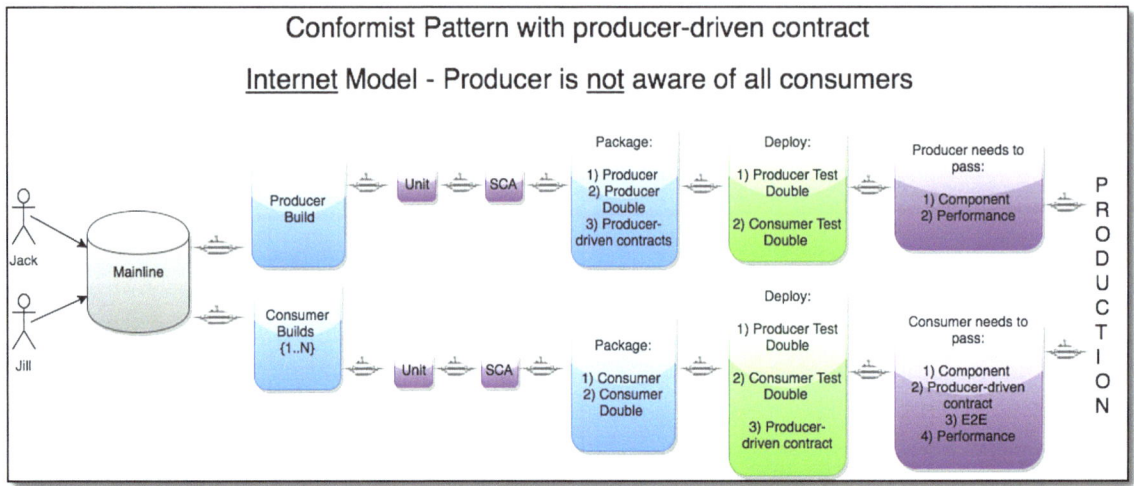

Figure 7 - Conformist

View, print and download from http://continuity.world

The Consumer Pipeline should:

- Build the code

- Pass unit tests

- Pass Static Code Analysis checks

- Publish a versioned artifact of the Consumer and its double

- Deploy the Producer's double, the Consumer's double and the Producer-driven contracts

- Pass component, contract, E2E integration and performance tests

- Go live!

"Conformist" is a preferred pattern amongst teams for the simple reason that less upfront agreements need to be sought.

Also, note that the Producer owns packaging and distribution of its double, and so does the Consumer. It is dangerous for the Producer to manufacture a Consumer's double, since the Producer may be unfamiliar with the Consumer's latest changes, and vice versa. This could lead to false positives or false negatives, thereby leading teams to question the integrity of the Pipeline.

Open Host Service, Published Language

The Pipeline could support multiple clients through the Open Host Service formalized as Published Language.

Vendors establish DSL (Domain Specific Language) that provides building blocks to implement Continuous Delivery Pipelines. DSLs should closely reflect ubiquitous language established as part of the Continuous Delivery domain model.

Pipeline DSL helps abstract away gory construction details and lowers the TCO (Total Cost of Ownership) to build and maintain Pipelines. We should invest into using DSLs since it frees us to focus on building applications that wow the world, instead of being bogged down by infrastructure.

DSLs could require us to write code, scripts and/or configuration in specified formats.

Common formats are:

1. Jenkinsfile (Groovy-ish) in Jenkins 2.0 Pipeline-As-Code
2. YAML in Travis CI
3. JSON in Concourse CI
4. YAML in AWS Code Pipeline
5. XML templates in GoCD
6. Etc.

DSLs are versioned in source control repository and hence are fully traceable. A Pipeline is a product and needs to be auditable, and the DSL enables us to apply the same gold standards to the Pipeline as the products that flow through the Pipeline.

Anticorruption Layer

A Pipeline doesn't live in an island of its own and hence it is important to translate and insulate its interfaces with external systems using ACLs or anticorruption layers.

To release software from source code repository to Production, the Pipeline needs to interact with systems that are:

- Internally owned-and-operated by the organization

- External SaaS (Software as a service), IaaS (Infrastructure as a service) and PaaS (Platform as a service) vendors

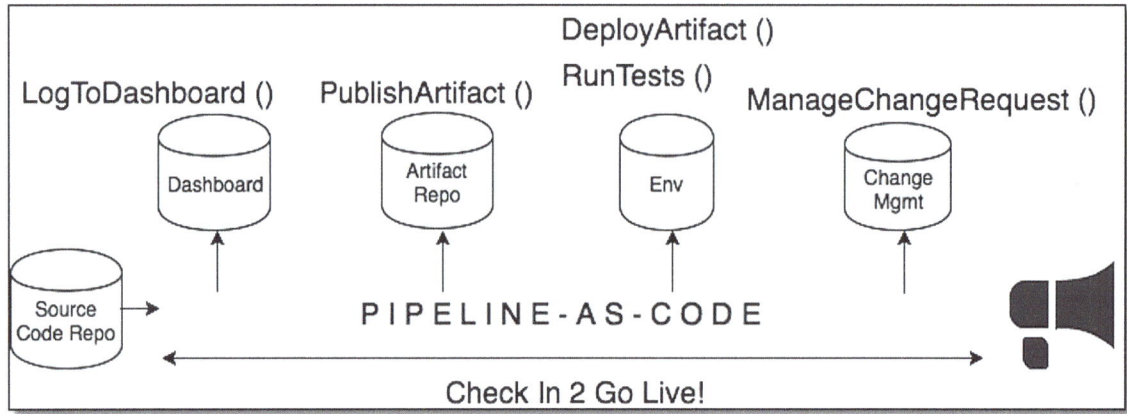

Figure 8 - Anticorruption Layer

View, print and download from http://continuity.world

Similarly, the organization could be deploying into on-prem data center VMWare VMs and could be migrating to AWS (Amazon Web Services) EC2 instances or Microsoft Azure or Pivotal's Cloud Foundry Or Heroku, while honoring the DeployArtifact () interface with the Pipeline.

Let's consider some examples where anticorruption layers (ACLs) could save the day:

- The Pipeline interacts with a source control repository.

If this repository changed from an on-prem installation to Atlassian's Bitbucket on the Cloud or AWS CodeCommit, it should be business as usual for the Scrum Teams who are customers of the Continuous Delivery Pipeline.

- The Pipeline interacts with a dashboard to log KPIs (Key Performance Indicators) that enable the leadership team to make informed decisions.

  Whether the dashboard is an on-prem ELK (Elastic, LogStash, Kibana) installation or * as a service providers like SumoLogic or AWS CloudWatch, there could be a LogToDashboard () interface that expects, as arguments, the endpoint and the message to be logged.

- The Pipeline interacts with an artifact repository to publish versioned artifacts.

  Whether the repository is Sonatype's Nexus Repository or JFrog's Artifactory or Amazon's S3, there could be a PublishArtifact () interface that expects, as arguments, the artifact and the endpoint where it should be published.

For external collaboration to be robust, good modularity is a must. Compare this to the Façade software design pattern that provides a simplified interface to a body of code.

Big Ball Of Mud (And Tests)

While studying the domain model for Continuous Delivery Pipelines, we looked at:

- Loosely coupled components (the smallest distributable and testable units) making up subsystems (the smallest deployable and runnable units), and

- Loosely coupled subsystems (the smallest deployable and runnable units) getting assembled into a system or several mini systems.

Assembly or composition leads to a monolithic architecture that's often referred to as a "Big Ball Of Mud". When a big ball of mud rolls down a hill, it collects dirt on its way and this is typically the case when a monolith is released from Dev to Stage to Production.

A corollary: A big ball of mud results in a big ball of tests. Certification of a monolith necessitates a highly integrated E2E (End-to-End) environment that adds exponential levels of difficulty in achieving Continuous Delivery in enterprises.

A big ball of mud (and tests) should be strangulated gradually over time to SOA (Service Oriented Architecture) or microservices architecture that offer better modularization. This enables parts of the code, that have a higher rate-of-change than others, to be released faster. Strangulation needs to happen gradually since the cost-benefit analysis doesn't always add up.

The following issues should be carefully considered while making architectural enhancements:

- Big bang architectural changes are incredibly hard to execute, and vision without execution is no good.

- Re-architecture isn't the top priority for most organizations, since the business needs to keep running with the same, limited set of resources.

- RoI (Return of Investment) may be high for partial strangulation of monoliths into microservices, while complete strangulation may need to be postponed for further assessment.

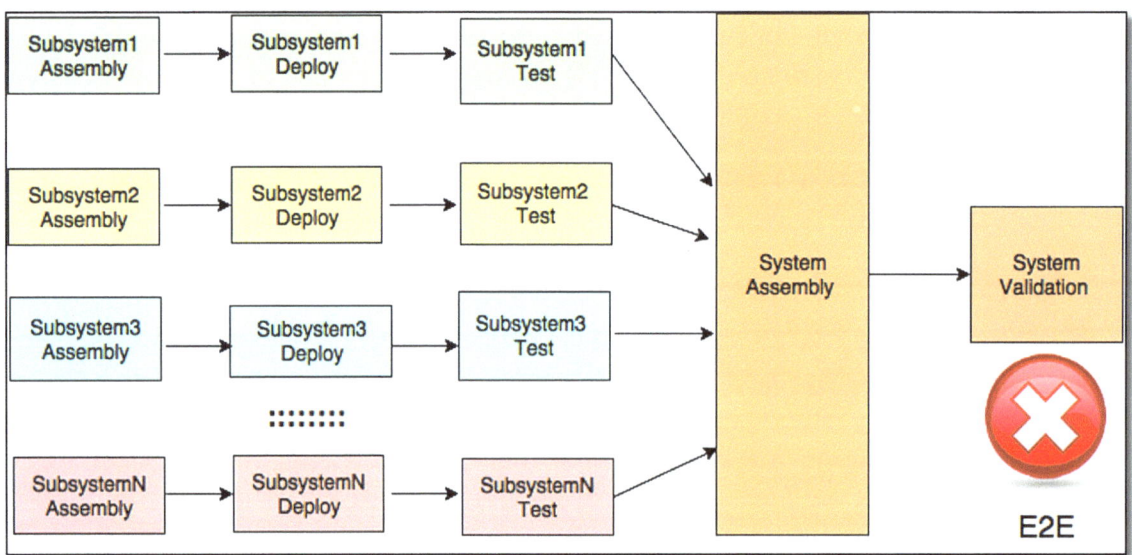

Figure 9 - Product Composition Anti-pattern

View, print and download from http://continuity.world

Figure 9 illustrates subsystems {1...N} being assembled into one system. We should question this composition anti-pattern and the need for an E2E (End-to-End) environment.

A big ball of mud (and tests) is strongly discouraged since we notice that faster teams move at the pace of the slower teams, which reminds me of the timeless saying: "A chain is only as strong as its weakest link!". However, we may judiciously need to leave the last piece of (the now much smaller) monolith alone by itself.

Separate Ways

If a team can get a bounded context that is independent of other contexts, then it is usually best to maintain that isolation.

If we have a single independently deployable artifact that can make its way to Production without requiring heavy-duty integration tests and/or E2E (End-2-End) environments, it is due to decoupled architecture that will stand us in good stead.

Bounded contexts in isolation are free to evolve on their own and thus do not incur the overhead of a Context Map or a Shared Kernel or other patterns that deal with interdependence between models.

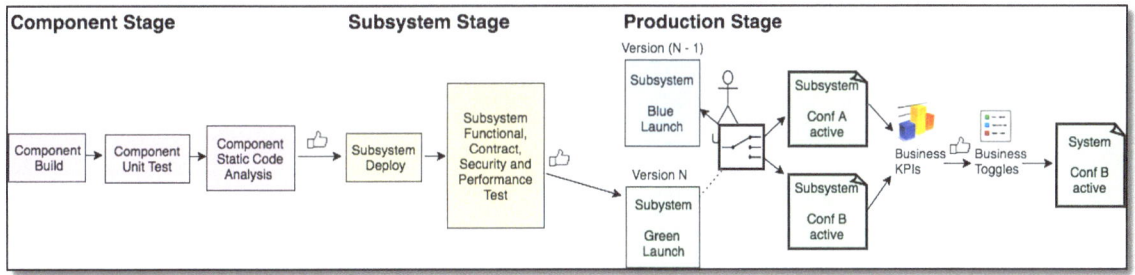

Figure 10 - Independently Deployable Artifact

View, print and download from http://continuity.world

The independently deployable artifact may still require integration tests with other independently deployable artifacts. In a SOA (Service Oriented Architecture) or microservices world, integration could become trickier with heightened levels of granularity.

We could (and should) use mocks and stubs for neighboring services and in-memory backing stores to improve velocity in the Subsystem Phase. Moreover, since each artifact could be deployed at its own pace, forward and backward compatibility between the artifacts should be a design requirement.

KEY TAKEAWAYS FROM CHAPTER 3

- While a unified domain model is great, it is at best aspirational in large enterprises where different teams work on different parts of the model.

- Product owners and architects should make it a top priority to maintain the integrity of the model, so that the core pieces can stay unified.

- Context Map is used to map the entire terrain. The technical leader who understands the entire lay of the land is often times the best person to run the Scrum of Scrums.

Chapter 4 | Continuous Analytics And Insights

Teams are required to *zoom in* and solve hard problems every day. They need to tease offensive incidents apart to analyze the root cause, and ensure that those incidents never happen again.

While it is paramount to be able to get down on our knees and dig in, it is just as important to be able to *zoom out* and fathom the big picture. We need to see the forest through the trees.

We should emphasize on learning patterns and trends over a period of time that offer direct insights into our business. These help us make tough calls that pay rich dividends in the long run.

Organizational KPIs Vs. Departmental KPIs

What if *one* department in an organization feels that it has met its goals? May or may not mean that the organization has met its goals.

What if *all* departments in that organization feel that they have met their goals? This could lead to a success story for the organization; however, there are cases to prove otherwise. How is this possible? Departmental KPIs (Key Performance Indicators) could have bias introduced by that department's vested interests, especially when there is departmental infighting. KPIs instituted by Dev, by QA, by Release and by Operations are great, however, sometimes they don't add up to propel the whole organization forward.

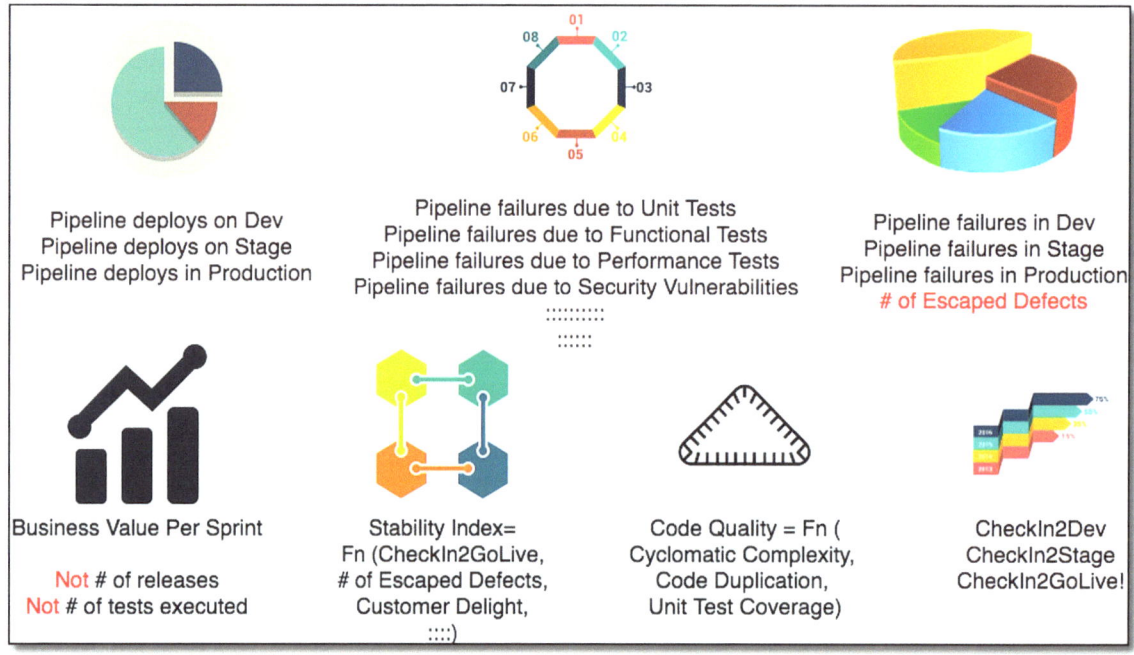

Figure 11 - Key Performance Indicators

View, print and download from http://continuity.world

We can counter this bias by establishing KPIs at an organizational level, or better still, reduce/eliminate departments or silos by internalizing the DevSecOps philosophy.

Some examples of KPIs that lead to organizational insights are illustrated in Figure 11. These will give you an idea of what to measure and how, so that you can get a pulse of your business by taking one look at the organization's dashboard.

% Of Deployments To Dev Vs. Stage Vs. Prod

Let's say, out of the total number of deployments to Dev, Stage and Production in our organization last month:

- 70% of all attempted deployments were only on the Dev environment,
- 50% of Dev deployments made it to the Stage environment, and
- 10% of Stage deployments made it to Production.

Assumption: This organization has three environments for product releases – Dev, Stage and Production. We could have fewer or more, and this concept would still be relevant. Something to remember would be that more environments have no correlation to better quality and on the contrary could add to maintenance overhead.

Instead of getting judgmental about the individual data points (70%, 50% and 10%), we need to focus on trends. Let's say, this was the data in the first month, and the data for the second and the third months were in that same ballpark. So, the organization has an established pattern in the first quarter.

However, much to everyone's surprise (or shock), the fourth month revealed that the organization failed to deliver anything to the customer:

- 80% of all attempted deployments were only on the Dev environment,
- 20% of Dev deployments made it to the Stage environment, and
- 0% of Stage deployments made it to Production.

There is no obvious conclusion as to why most of the effort was spent in Dev. However, several questions come to mind:

- Did the requirements change too often, as in, was Marketing/Business interrupting the Sprints frequently, so much so that the Engineering team was always playing catch-up?

- Didn't the teams see business value of frequent deployments to Production? Do they understand how they contribute directly to the top line and to the bottom line? Do they need training?

- Was the Stage environment unstable, such that the Pipeline kept aborting after Dev?

Turned out, that in that fourth month, this team had added solid performance tests for the very first time and had integrated them with the Pipeline. They then discovered that the Dev environment was way inferior to either Stage or Production and hence the integrity of the performance tests was questionable on Dev. So, they went through an intensive exercise to move away from on-prem solutions onto the Cloud, where their Dev, Stage and Production were (almost) identical.

Soon enough the Continuous Delivery Pipeline was catching authentic performance issues, and was preventing sluggish code from reaching Production. Further digging revealed that this team had severe performance issues in Production in the past, and sure enough, in the next few months and quarters, they started to deliver quality products frequently and predictably to their customers.

% Of Failures Per Test Type

Typically, the following kinds of tests are integrated with the Pipeline:

- Unit Tests

- Static Code Analysis to detect deviations from coding best practices

- Functional Tests

- Integration Tests

- Performance Tests

    o Under normal conditions

    o Under extreme conditions

- Application Security Tests

    - SAST - Static Analysis Security Testing to detect security vulnerabilities

    - DAST - Dynamic Analysis Security Testing

    - Penetration Tests

For Team A, over the last quarter, 10% of the Pipeline test failures were due to Unit tests, 10% due to Functional tests, 30% due to Integration tests, 25% due to Performance tests and 25% due to Security tests (SAST, DAST and Penetration).

For Team B, over the same quarter, 30% of the Pipeline test failures were due to Functional tests, 50% due to Integration tests and 20% due to Performance tests.

Several questions come to mind:

- Team B did have many unit tests, so why were there no failures? Tests that always pass aren't always good news. Did they find unfair methods to have unit tests pass all the time? We are painfully aware of how easy that is.

- Is it possible that both teams write excellent quality code that's approved by their Static Analyzers?

- Does Team B have more integration touch points with other teams than Team A?

- Does Team B do security tests manually, or worse, skip them?

Code reviews revealed that Team B indeed had resorted to unfair practices to pass unit tests. Which then led us to:

- Reverting penalties set by inexperienced management for failed unit tests, and

- Relaxing goals to achieve 100% unit test coverage.

Sprint retrospectives revealed that both teams didn't have faith in their current Static Code Analyzer, and hence refrained from failing Pipelines when the static analyzers reported errors. Which then led us to do more proof of concepts with smarter industry-standard code analyzing tools.

Team B didn't necessarily have more integration points than Team A, however, they talked to an Oracle database that was never considered to be a subsystem and wasn't rolled into the Continuous Delivery Pipeline. The Oracle DBAs managed database releases manually, and consequently there was human coordination overhead, which led to errors and frustration. Which then led us to prioritize automated database change management and Continuous Delivery for the database.

% Of Defects Discovered Vs. Escaped

Budget for the Continuous Delivery Pipeline cannot be shoved under the carpet and hence proving RoI (Return on Investment) is essential. Some of us have gone through what I call "Show Me The Money!" moments, where we demonstrate the cost-benefit analysis of building (and maintaining) Pipelines.

% of defects discovered by the Pipeline is a *positive* return on building Pipelines and can be further decomposed into:

- % of defects discovered in Dev,
- % of defects discovered in Stage, and
- % of defects discovered in Production.

% of escaped defects are defects that escaped the Pipeline, and is a *negative* return on investing into Pipelines. Escaped defects cause downtime and customer dissatisfaction.

Escaped defects can be due to:

- A missing DoD (Definition of Done), leading to teams "completing" Sprints that are incomplete
- Low test coverage, leading to dysfunctional customer use cases
- Low automation, leading to human errors and manual coordination overhead
- Insufficient KPIs leading to poor gating criteria for code promotion from Dev to Stage to Production

- Misunderstandings between Marketing and Product on customer expectations

- Misunderstandings between Product and Engineering on vision and priorities

- Misunderstandings between Development, QA, Release, InfoSec and Operations on operational excellence requirements

- Etc.

The Pipeline, by surfacing these issues, helps the organization prioritize and pay off its technical debt. Also, note that these defects could be due to test failures, bad configuration, erroneous deployments, environmental problems and data issues.

If done right, Continuous Delivery Pipelines improve culture, processes, trust, velocity and productivity in an organization, and positive returns on investment are normal. It is worthwhile to note that intensive upfront investments in automation can sometimes cause nervousness amongst stakeholders but patience eventually pays off for those who do the right thing, even when no one is looking.

Business Value Delivered Per Sprint

A flawed measure of velocity is the number of story points played in a Sprint.

Number of points attached to epics, user stories, features and tasks played out per Sprint is a quantitative measure and yet is subject to human perceptions. We could refine the definitions of epic, user story, feature and task within our teams, and still get baffled by the variations in opinion and granularity.

Number of story points for a feature is influenced by the Product Owner's style and ability to disintegrate large problems into smaller sub-problems. Teams tend to be safe than sorry for high visibility projects, given the penalties imposed for execution delays. Higher numbers could be thrown out as points based off people's perceptions of how complex a given story is.

Contrast number of story points to business value delivered per Sprint. A business goal is objective and straightforward, like:

- % of online advertisement views that result in advertisement clicks per month

- 20% increment in page views per week

- 50% reduction in page load time

- Revenue increase of 10% per quarter

Another flawed measure of velocity is the number of Production releases per Sprint. While this reflects Time2Market, moving bits from Point A to Point B may or may not mean much for the organization, unless business value is realized.

No hand waiving should be allowed in articulating business KPIs, and it should be measurable in clear terms. If we trend business KPIs across Sprints, months, quarters and years, we should be able to tell with certainty whether we are rowing in the right direction or if we need to make adjustments.

Weighted Stability Index

The stability of a system is typically represented by its uptime (or downtime), and is a key metric that impacts our customers. Blind focus on uptime (or downtime) could make us risk aversive and innovation could suffer. This could also lead to friction between Development, QA, Release, InfoSec and Operations.

It's best that our organization collectively determines what "stability" means. It is prudent to throw in a few more KPIs to this mix, such that the Stability Index avoids skewing on one specific metric, and is more a function of several relevant ones. Moreover, different metrics could have different weights, since they could have varying impacts on the overall index.

Here's a sample, with suggested weights as percentages, which you can tune according to your needs:

Stability Index = Function of [

- 30% (Check-In-2-Go-Live)

- 30% (# of escaped defects)

- 40% (customer delight)

], where

- Check-In-2-Go-Live is the time it takes for a check-in to go live in the hands of our customer. This influences the MTTR (Mean Time To Repair/Resolve) of escaped defects.

- # of escaped defects represents issues that the Pipeline was unable to detect with the gating criteria for code promotion from Dev to Stage to Production.

- Customer delight (or the lack of it) is subjective, and is harder to quantify. However, this could be a function of:

    a. % of repeat customers YoY (Year over Year)

    b. # of functional, performance and security defects that are logged by the customer, even if this was a misunderstanding that wasn't clarified in the User Manual

    c. # of user experience degradations that are logged by the customer. For example, the increment in the number of clicks required to complete a transaction from the last release to this could cause customer dissatisfaction.

Trending Stability Index over time would reveal if we are rowing in the right direction, or if need a course correction. The teams need to come up with their own definition of "stability", and more importantly fine-tune the function and the weights based on their discoveries.

Weighted Code Quality Index

The quality of the code we write determines the quality of the applications that are delivered to our customers. "Garbage in, garbage out" is a known condition that will cause pain unless addressed with urgency.

Rather than inspecting the quality and security *after* the application is built, DevSecOps principles advocate building quality and security into the application *before*:

- The product is finished, and

- The bugs become elements of the design.

Prevent security from being an afterthought that a separate InfoSec team needs to inspect and fix. Security is not an inspection point, since it should be baked into the design.

This is a sample Code Quality function with suggested weights as percentages, which you can tune according to your needs:

Code Quality = Function of [

- 20% (Cyclomatic complexity)
- 20% (Code duplication)
- 10% (Unit test coverage)
- 10% (Static code analysis to detect violations in coding best practices)
- 20% (Static code analysis to detect security vulnerabilities)
- 20% (Static code analysis to detect memory leaks)

], where

- Cyclomatic complexity helps us understand if there are too many code paths that hinder testing and code reviews
- Code duplication helps us understand if we need to refactor and modularize our code instead of copying and pasting carelessly (humorously referred to as the "C&V Technology")
- Unit test coverage helps us understand if we have sufficient coverage as the first line of defense. A 100% code coverage requirement is the easiest thing to mandate, however, there's enough data to suggest that this metric can be gamed. Additionally, this metric by itself doesn't mean much for our customers. So, doing it is fine, however, overdoing it is not.

- Static code analysis to detect deviations from coding best practices helps us understand if we have good hygiene in our code

- Static analyzers help us detect security vulnerabilities early on that would otherwise pose security threats to the system and the organization

- Static analyzers can indicate potential memory leaks that would eventually cripple the system

Overall, DevSecOps is about embedding Development, Security and Operations principles into our product design, irrespective of how teams are structured. Teams need to come up with their definition of good quality code, and then fine-tune the function and the weights based on their discoveries.

Check-In-2-Go-Live With # Of Escaped Defects

Continuous Delivery Pipelines should bring speed and predictability without compromising quality. By speed, we mean responsible speed and not suicidal speed.

The simplest way to measure this would be to understand how Check-In-2-Go-Live trends over time with respect to the # of escaped defects.

As a refresher,

- Check-In-2-Go-Live is the time it takes for a check-in to go live in the hands of our customer. This influences the MTTR (Mean Time To Repair/Resolve) of escaped defects.

- # of escaped defects represents issues that the Pipeline is unable to detect with the gating criteria, when it promotes code from Dev to Stage to Production.

For example, if Check-In-2-Go-Live decreases from 2 weeks to 2 hours and during the same time window, the # of escaped defects goes up by 50%, we need to make adjustments to the software gates that promote code from Dev to Stage to Production. See more on software gates in chapter "Segregation/Separation Of Duties".

Check-In-2-Go-Live is a rough summation of Check-In-2-Dev, Dev-2-Stage and Stage-2-Production, along with the wait times at the manual gate(s), if we have any. This assumes there are three environments in the organization - Dev, Stage and Production - and we could apply this same principle to all environments that we have.

Concept-2-Cash

Before we focus our resources in optimizing metrics like Check-In-2-Go-Live, it is critical to understand where the bottlenecks are in the grand scheme of things.

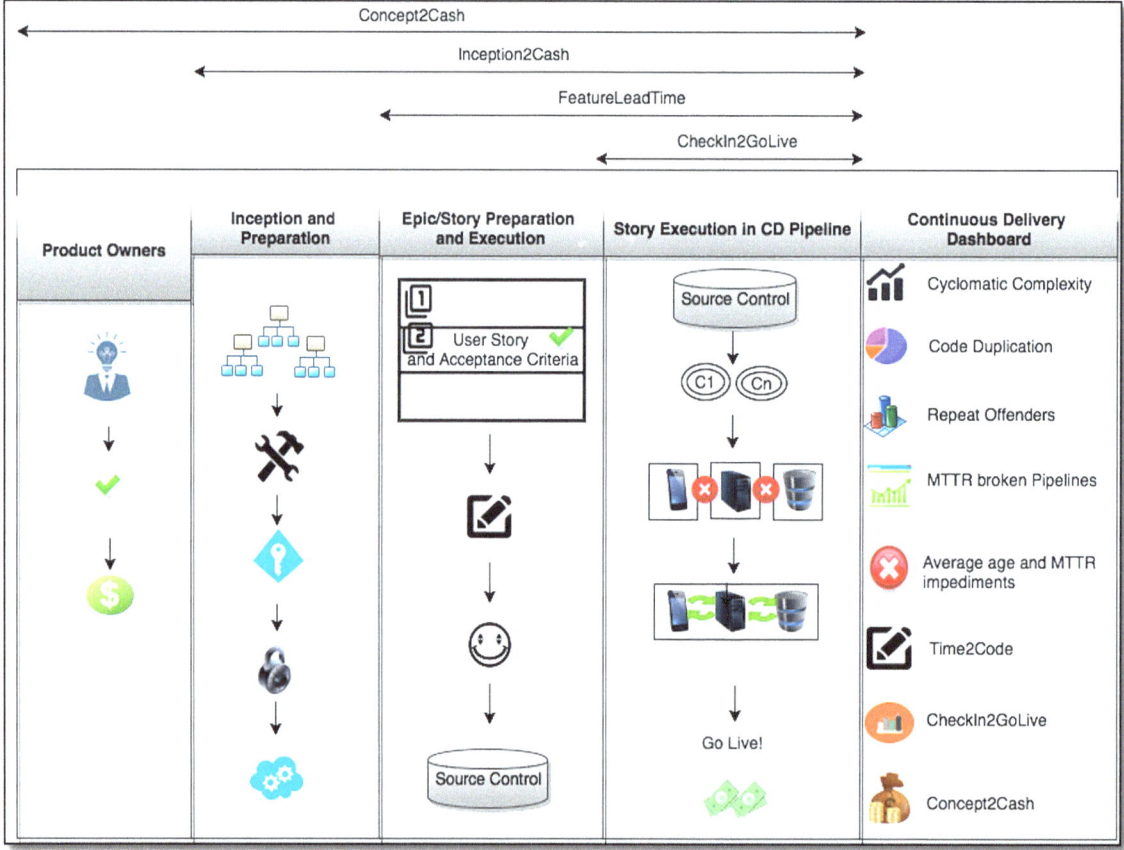

Figure 12 – Concept-2-Cash

View, print and download from http://continuity.world

Check-In-2-Go-Live is at the heart of a larger set of KPIs that the industry cares about deeply. Check-In-2-Go-Live affects FeatureLeadTime, which in turn affects Inception-2-Cash, which in turn affects Concept-2-Cash.

Enterprises lose valuable momentum in jumpstarting a new idea, while sifting through paperwork related to budget and approvals. While it is necessary to prove that a new idea is worth an investment, it is ironical that some organizations tend to waste millions trying to prove that an idea is worth millions. They have "experts" debating in rooms for months, who cost more than what it would cost a few of engineers to whip up a proof of concept that either proves or disproves a set of hypotheses. An occasional leap of faith is needed to get the prototype started without drama. The most practical way to mitigate risk is to establish periodic milestones and a clear DoD (Definition of Done) that the team understands and feels proud of.

In summary, we need to optimize Check-In-2-Go-Live for sure, and that is very much in the purview of the Continuous Delivery Pipeline. However, Figure 12 illustrates the other KPIs that organizations should optimize, like FeatureLeadTime, Inception-2-Cash, and Concept-2-Cash, which cannot be solved for within the Pipeline.

KEY TAKEAWAYS FROM CHAPTER 4

- Make informed decisions based off facts driven by KPIs. Avoid opinions.

- Banking heavily on one single KPI can lead to insights that are skewed on that one metric. Have a bunch of relevant KPIs create tension such that there is a center of gravity, which then serves as the guiding force.

- See the forest through the trees. Individual data points speak of isolated incidents that could be relatively harmless compared to repetitive offending patterns that only trends can reveal.

CHAPTER 5 | A DEVSECOPS SEED BACKLOG

A single prioritized product backlog is a key driver to an organization's growth and sustainability. A traditional Product Owner leans heavily towards development and quality, and sometimes overlooks the security and operational aspects of releasing products to their customers. There could be Product Owners in the Security and Operations teams, who are entrusted to fill in this void; however, multiple Product Owners tend to come up with separate copies of a "single prioritized backlog", thereby leading to fragmented and frictional execution.

This chapter enables Product Owners to address epics related to architecture, feature development, builds, tests, configuration, deployment, monitoring, KPIs, operational excellence, network, security and related issues that help teams experiment safely and turn great ideas into products. More importantly, smart designs help teams recover fast from not-so-great ideas.

This, by no means, is an exhaustive backlog and serves as a "Seed Backlog" to jumpstart teams. The following epics are in no particular order of priority, and you should prioritize these, along with other epics you have written, as per your business needs.

The following epics are categorized for Tech Leads, Engineers and Product Owners, based on who is most likely going to own these. However, irrespective of titles, ownership can lie with anyone with the appropriate skills.

For Tech Leads

Epic 1

As a tech lead
I want to measure the number of Production issues and MTTR (Mean Time To Resolve) Production issues

- *Before* building a Pipeline, to establish a baseline over which RoI (Return on Investment) will be calculated

- *After* building a Pipeline, to measure if percentage of Production issues per month/quarter has decreased

So that I know that the Pipeline provides responsible speed to my team

We should be able to trend the number of Production issues in an automated fashion. Manual computations can introduce errors and human bias.

If our MVP (Minimum Viable Product) for Pipeline involves a brand new greenfield application, we can consider doing a baseline analysis with an existing comparable product.

Epic 2

As a tech lead
I want to measure "Check-In-2-Go-Live", which is the time taken by a check-in to go live in Production in the hands of our customer

- *Before* building a Pipeline, to establish a baseline over which RoI (Return on Investment) will be calculated

- *After* building a Pipeline, to measure if velocity has increased

So that I can validate if this aligns with delivery speed improvement goals of my team

We should be able to trend the increment in velocity in an automated fashion. Manual computations can introduce errors and human bias.

If our MVP (Minimum Viable Product) for Pipeline involves a brand new greenfield application, we can consider doing a baseline analysis with an existing comparable product.

Check-In-2-Go-Live is at the heart of several metrics like FeatureLeadTime, Inception-2-Cash and Concept-2-Cash, and is a rudimentary reflection of velocity. Refer to section "Concept-2-Cash" in chapter "Continuous Analytics And Insights" for more details.

Epic 3

As a tech lead
I want to know what percentage of my tests are manual and are executed outside the Pipeline
So that I can accurately assess RoI (Return on Investment) of my Pipeline

Sometimes, teams tend to report tests that are integrated with the Pipeline as the total number of tests they have, and hence the "Percentage Of Automated Tests" always comes to 100%. If we have 10 tests, out of which 8 are integrated with the Pipeline and 2 are manually executed outside the Pipeline, we should report "Percentage Of Automated Tests" as 80%.

Epic 4

As a tech lead
I would like to define a test strategy around:

- Unit Tests

- SCA (Static Code Analysis) to detect deviations from coding best practices

- SAST (Static Analysis Security Testing) to detect security vulnerabilities

- Functional tests, involving components and contracts

- Integration tests, involving gateway and persistence

- Performance tests, and

- DAST (Dynamic Analysis Security Testing)

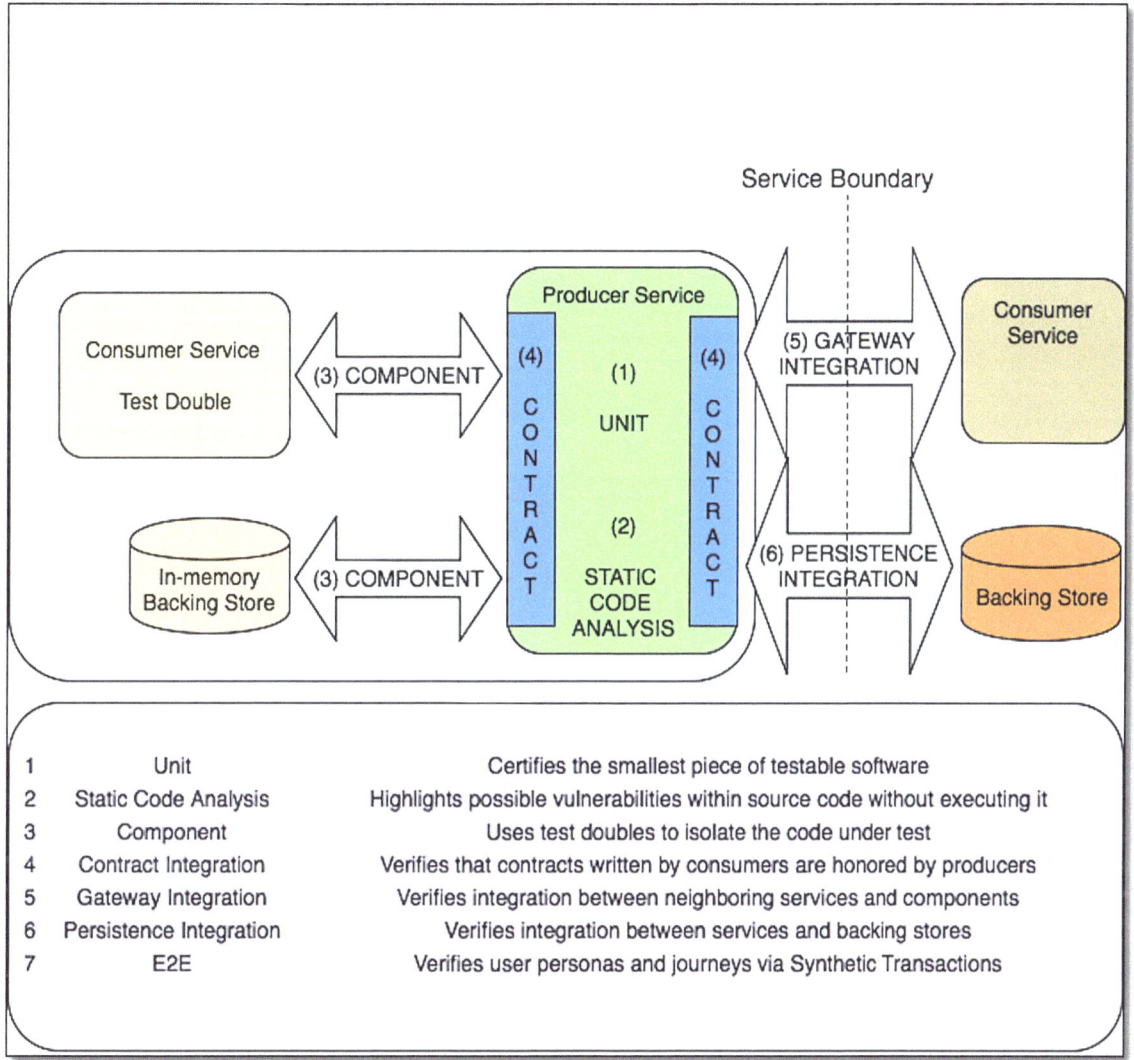

Figure 13 - Test Architecture For Microservices

View, print and download from http://continuity.world

So that I can take full ownership of validating my changes from Dev to Stage to Production

Figure 13 illustrates a use case where a Producer service communicates with a Consumer service and a backing store. It demonstrates the use of test doubles and an in-memory backing store in the Subsystem Phase of the Continuous Delivery Pipeline, while using the real service and the real backing store in the System Phase.

Refer to section "Domain Model For Continuous Delivery Pipeline" in chapter "Continuous Delivery As A Domain".

Epic 5

As a tech lead
I want to see teams negotiate *traceable* service contracts with the Pipeline
So that there's transparency in the:

- Automated gating criteria to promote code from Dev to Stage to Production

- Conditions under which those criteria get re-negotiated, and by who

- Conditions under which negotiated criteria get compromised, and by who

Epic 6

As a tech lead
I want to disintegrate my product architecture into:

- Components (the smallest distributable and testable units)

- Subsystems (the smallest deployable and runnable units that comprise loosely coupled components), and

- System (that comprises loosely coupled subsystems)

So that there's transparency over:

- The different phases of the Pipeline

- What versioned artifacts are produced in each phase

- What tests are executed in each phase to certify the artifacts produced

- What kind of software gates are used to promote code from one phase of the Pipeline to the next

Refer to section "Domain Model For Continuous Delivery Pipeline" in chapter "Continuous Delivery As A Domain".

Epic 7

As a tech lead
I want test statistics [Status: (Pass | Fail | Disabled | Incomplete), Duration] in the Pipeline
So that my test results can be exposed to compute Continuous Delivery analytics and insights, which would help our organization make informed decisions

The Pipeline should execute the following kinds of tests, and trend test results over time.

- Unit Tests

- Static Code Analysis to detect deviations from coding best practices

- Functional Tests

- Integration Tests

- Performance Tests

    o Under normal conditions

    o Under extreme conditions

- Application Security

    o SAST - Static Analysis Security Testing

    o DAST - Dynamic Analysis Security Testing

    o Penetration Tests

Executing all tests and aborting the Pipeline when gates are violated should be blocking (or, synchronous) calls. Publishing test results could be a non-blocking (or, asynchronous) call or a background process, especially if we experience network latency that slows the Pipeline down.

Disabled and incomplete tests should be treated as failed tests, as far as the software gates are concerned. If gates are violated, Pipelines should abort with appropriate notifications sent to the potential culprits and a distribution list. See more on software gates in chapter "Segregation/Separation Of Duties".

Epic 8

As a tech lead
I want to view Pipeline statistics [Status: (Pass | Fail | Disabled | Incomplete), Duration]
So that I can get factual indicators on how the Pipelines will impact my team's delivery speed

We should report the following stats in the Pipeline:

- Code commit(s) that triggered the Pipeline, along with the committer and commit IDs

- Pipeline status (Running | Pass | Fail | Disabled | Incomplete)

- Pipeline's total duration, i.e., Check-In-2-Go-Live, along with duration of each phase of the Pipeline, i.e. Check-In-2-Dev, Check-In-2-Stage, and Check-In-2-Prod, assuming that there are three milestones Dev, Stage and Production

- Test execution summary (Total # | Passed # | Failed # | Repeat Offenders #)

- Percentage of automated tests

Refer to chapter "Continuous Analytics And Insights" for details on organizational KPIs (Key Performance Indicators) that could be reported with the help of Pipelines.

Epic 9

As a tech lead
I want 99% uptime in my Pipeline and the systems that it interfaces with

So that I can do frequent and predictable releases through the Pipeline from Dev to Stage to Production

Some standard use cases where the Pipeline interacts with systems are:

- A source control repository like Atlassian's Bitbucket or GitHub

- A dashboard, like SumoLogic or ELK Stack, to log KPIs (Key Performance Indicators) that enables the leadership team to make informed decisions

- An artifact repository to publish versioned artifacts, like Sonatype's Nexus Repository or JFrog's Artifactory or Amazon's S3

- On-prem data center VMWare VMs or Amazon's AWS or Microsoft Azure or Pivotal's Cloud Foundry PaaS (Platform as a Service) or Heroku PaaS etc.

- Atlassian's Jira to manage bugs and change requests

- A configuration server to manage configuration

- A web-browser and mobile device farm to execute web and mobile functional tests, like SauceLabs or Amazon Device Farm

- An external SaaS service like BlazeMeter to run performance tests

- Etc.

Since external providers are involved, the uptime requirements from vendors need to be made part of their contractual obligations.

Epic 10

As a tech lead
I want SLAs (Service Level Agreements) defined for each phase of the Pipeline
So that we know the bottlenecks to increase delivery speed, quality and predictability

Since external providers are involved, SLAs from vendors need to be made part of their contractual obligations.

Epic 11

As a tech lead

I want to see performance benchmarks established that would decide pass/fail criteria of performance tests
So that I have high confidence in the performance tests that are integrated with the Pipeline

The Product Owner should set performance benchmarks based on customer expectations and those benchmarks should be used to fail or pass performance tests in the Pipeline.

Epic 12

As a tech lead
I want X months of test evidence retained for production artifacts
So that I can comply with regulatory and audit guidelines

History of faulty artifacts that have been delivered to Production by the Pipeline needs to be traced back for a certain number of months.

Some internal auditors require the Pipeline to attach an automatically generated test evidence document to a System of Record. Internal audit teams typically use change management systems like Jira to answer audit questions, however, this is a questionable practice since user stories can be manually signed off in Jira even when Pipelines fail.

Most auditors, especially external, are unfamiliar with the world of Pipelines. They could look at the Pipeline execution reports directly, which have rich information on test evidence.

Epic 13

As a tech lead
I want to participate in defining a rollback strategy
So that as the last resort, I can rollback to the last certified version of the product

We should try and roll forward in the true spirit of Continuous Delivery, however, it is advisable to have a rollback option available till we are disciplined enough.

Rollbacks can happen in Dev, Stage or Production, and are usually easier when the offending features can be turned off with toggles.

Epic 14

As a tech lead
I want to establish security and audit guidelines for the Pipeline
So that I can trust the Pipeline to deploy artifacts to Production safely

Most organizations do not take Pipeline audits seriously till they practice Continuous Delivery or Continuous Deployment to Production. We should treat Pipelines as Production assets and establish audit guidelines the same way we handle the products flowing through the Pipelines.

Internal and external auditors are not well versed with Pipelines and end up seeing Pipelines as a hindrance to implement Segregation/Separation of Duties. On the contrary, Pipelines generate first class audit trails and radically improve transparency of each and every action taken by each and every person and automated agent in the organization.

Epic 15

As a tech lead
I need tools to:

- Audit actions taken by headless users in the Pipeline

- Trace headless actions back to a named user account

So that I can debug issues that were caused by artifacts deployed by the Pipeline

Epic 16

As a tech lead
I want established metadata for Pipelines that disclose:

- Their location/URL

- The product that's flowing through them

- The Scrum team that is involved

- Single point of contact who has financial accountability and full responsibility

- The dashboard to which they log KPIs

- The automated agent(s) that are executing actions on behalf of humans

So that questions, if and when they arise, can be answered with ease

Epic 17

As a tech lead
I want Pipelines to treat PII (Personally Identifiable Information) with secure data management techniques
So that the Pipeline meets security standards

Aggregated data may not be considered PII. Company policies should clearly state the safeguards that need to be built into the Pipeline, and those should be incorporated into the audit guidelines for Pipelines.

Epic 18

As a tech lead
I want the Pipeline to treat financial data with secure data management techniques
So that the Pipeline meets security standards

If we inadvertently become party to financial data, we could be considered "insiders" and may need to abide by requirements that are more stringent than those for regular employees.

Epic 19

As a tech lead
I want to provide business toggles to Product Owners
So that they have flexibility to decide:

- When to release finished features, and

- Which market to target

Business toggles can warrant a user interface since business folks are the primary users of these toggles.

Epic 20

As a tech lead
I want to define and measure business KPIs for my product
So that I can determine:

- Whether my product meets its purpose

- Whether we should bear the cost to maintain it in Production

- Whether we should grow or shrink the team

The business KPIs for a product reflects what success looks like. For example:

- % of online advertisement views that result in advertisement clicks per month

- 20% increment in page views per week

- 50% reduction in page load time

- Revenue increase of 10% per quarter

We should instrument the Pipeline so that business value measurements can be automated and trended across Sprints, months, quarters and years. The trends will help us make informed decisions on team rewards, budget and priorities. Refer to chapter "Continuous Analytics And Insights" for more details.

Epic 21

As a tech lead
I want to retain the last X runs of the Pipeline in history
So that I can compare the current execution report to the previous ones to troubleshoot

We should be careful of how much history we preserve, since this adversely affects storage. Since storage has become inexpensive, we typically don't lose sweat over the increased storage cost. However, after months, quarters and years, this could come back and bite us, if we aren't mindful.

Epic 22

As a tech lead
I want to run multiple active configurations of my application at the same time
So that I can support A/B testing

The MVP (Minimum Viable Product) can start with two active configurations, and then increase to multiple at a later time.

A/B tests may be considered external to Continuous Delivery Pipelines, since they are executed manually. However, since Continuous Delivery allows at least one manual gate, this can be part of the Pipeline domain model.

Epic 23

As a tech lead
I want to use industry standard test frameworks that integrate well with the Pipeline's tool-chain
So that these frameworks write out test reports in standard formats, which can be parsed and graphed with ease

Test frameworks like JUnit, TestNG and Jasmine write out test reports in standard formats like JUnit. Moreover, reporting formats like JUnit and TAP (Test Anything Protocol) can be parsed and graphed by standard Pipeline orchestrators, thus giving a bigger bang for the buck.

For Engineers

Epic 1

As an engineer
I would like to spin up the Pipeline locally the same way it spins up in the Cloud (or on VMs in the Data Center, for that matter)

So that I have an identical local environment and can pre-empt problems that would otherwise occur later

Containerized Pipelines have the advantage of spinning up and down as needed on predictable, traceable and versioned images.

Epic 2

As an engineer
I would like service owners to build, maintain and distribute their respective test doubles
So that consumers of those services don't encounter false positives or false negatives caused by outdated test doubles

Epic 3

As an engineer
I want my code to be promoted automatically from Dev to Stage to Production through software gates (instead of manual gates)
So that I am able to build, configure, deploy, test and release software with speed, quality and predictability

See more on software gates in chapter "Segregation/Separation Of Duties".

Epic 4

As an engineer
I want a Pipeline visualizer
So that I can look at the entire Pipeline as one graph and diagnose the bottlenecks easily

Epic 5

As an engineer
I want to use the right versions of my dependent artifacts, tests and foundational pieces
So that I can reduce/eliminate false positives and false negatives

Epic 6

As an engineer
I want to see in the Pipeline Log:

- Individual Test Status (Pass | Fail | Disabled | Incomplete | Start Time | End Time | Duration | TraceLog)

- Test Summary (Total# Pass | Total# Fail | Total# Disabled | Total# Incomplete | Total Duration)

So that I can diagnose and fix test failures with ease and within the SLA (Service Level Agreement) of that phase of the Pipeline

Epic 7

As an engineer
I want my test suites to return zero for success and non-zero codes for failure
So that the Pipeline can decide whether to proceed or abort

Depending upon how we design the test suite, the non-zero return code could:

- Be 1, signifying any kind of failure that requires the Pipeline to abort.

- Signify the total number of test failures. For example, a return code of 5 could imply that 5 tests failed, without indicating the type of test or the phase of the Pipeline the failure happened at.

- Indicate the type of failure. For example, 1 indicates a unit test failure, 2 indicates a static code analysis failure, 3 indicates a functional test failure, 4 indicates a performance test failure etc.

There are a couple of ways we could induce the Pipeline to abort on failures:

- Exactly at the first failure. In this case, the Pipeline aborts immediately at the first failure it encounters.

- When there's at least one failure. In this case, the Pipeline is allowed to run it's entire race and report all failures till the point it was completely crippled.

Note that the return codes reflect the nature in which the failures are being induced.

If we deploy to Production with failed tests and open bugs, it means we don't trust those tests. It is inefficient to waste cycles in bug scrubs prioritizing and de-prioritizing bugs.

Also, disabled or incomplete tests should be reported as failures and the Pipeline should abort and call for human intervention.

Like any other product, the Pipeline needs to be validated before we can release products through it. Ensure that an incomplete/failed/disabled test aborts the Pipeline. In a subsequent step, fix the problem and demonstrate that the Pipeline starts to flow again.

Epic 8

As an engineer
I want to use single-responsibility headless users
So that an audit can accurately establish ownership and accountability of Pipeline actions

Headless users are automated agents, which perform Pipeline actions, like:

- Checking out code from source control repository

- Publishing to the artifact repository, like Sonatype's Nexus Repository or JFrog's Artifactory or Amazon's S3

- Deploying to an on-prem VMWare VM or AWS (Amazon Web Services) or Microsoft Azure or Pivotal's Cloud Foundry etc.

- Executing tests and collecting test results

- Etc.

When diagnosing a Production defect, we should easily identify the headless user and which team it represents. Subsequently, we should trace the offending action back to the root cause and a named user who is accountable.

Epic 9

As an engineer

I want to safely publish certified artifacts to the artifact repository with the right credentials
So that these artifacts are securely available for distribution and downstream consumption

Epic 10

As an engineer
I want to be able to have a Dev environment that's identical to Stage and Production
So that I don't run into environmental issues that I can't reproduce on Dev

Epic 11

As an engineer
I want to have traceable Pipeline code and configuration
So that I can troubleshoot Pipeline failures the same way I troubleshoot any other application

We should be able to trace Pipeline changes in source control repository to understand:

- Who made the change

- What the change was

- When the change was made

This will be needed during pipeline audits, especially for organizations that are practicing Continuous Delivery and Continuous Deployment to Production.

Epic 12

As an engineer
I want at least Unit Tests and Static Code Analysis to certify components (the smallest distributable and testable units)
So that I have high confidence in the Component Phase of the Pipeline

The Static Code Analysis should include detection of security vulnerabilities, possible memory leaks, deviations from coding standards etc.

Refer to section "Domain Model For Continuous Delivery Pipeline" in chapter "Continuous Delivery As A Domain".

Epic 13

As an engineer
I want functional, security and performance tests to certify isolated subsystems with neighboring subsystems mocked
So that I have high confidence in the Subsystem Phase of the Pipeline

Refer to section "Domain Model For Continuous Delivery Pipeline" in chapter "Continuous Delivery As A Domain".

Epic 14

As an engineer
I want integration and performance tests to be executed to certify the system
So that I have high confidence in the System Phase of the Pipeline

Integration tests include checks and balances for:

- Inter-system integration

- Intra-system integration

- Backward compatibility

- Forward compatibility

- Interfaces

- Network

Avoid System Stage in your Pipeline. However, enterprises with legacy code and highly coupled product architecture have to entertain it. Refer to section "Domain Model For Continuous Delivery Pipeline" in chapter "Continuous Delivery As A Domain".

Epic 15

As an engineer
I want to protect secrets, certificates and credentials with secure key management techniques
So that the application passing through the Pipeline meets security, audit and compliance standards

Epic 16

As an engineer
I want to see Static Code Analysis reports published to a persistence layer
So that false positives can be marked off

Epic 17

As an engineer
I want to institute hooks in my deployment automation
So that I can:

- Perform prerequisites in the *pre-deploy* hook

- Perform validations in the *post-deploy* hook, for example, run a health check or execute a smoke test

If the smoke test in the post-deploy hook fails, extensive functional, integration, security and performance test suite execution can be postponed.

Compartmentalizing failures into 3 buckets - pre-deploy, deploy or post-deploy – help notifications to be sent to the right distribution channels. It is possible that a person who typically fixes configuration management issues is different from the team member who debugs functional test failures.

Epic 18

As an engineer
I want to follow a standardized directory structure in source control repository
So that I:

- Have well-organized code, configuration, tests, and data

- Can institute rules on triggering the Pipeline

- Can institute rules on avoiding triggering the Pipeline based off commits to particular directories that have no bearing on the Pipeline

Epic 19

As an engineer
I need infrastructure that supports our traffic load under:

- Normal conditions, and

- Extreme stress

So that the Pipeline can execute performance tests that simulate real life conditions

Epic 20

As an engineer
I want to design and implement feature toggles - build-time as well as run-time
So that:

- Unfinished features can be made dormant as and when required by the team

- Automated rollbacks can be performed in Production to stop the bleeding, while the team works on the fix

Moving the bits isn't the same as turning them on. An engineer could fall sick in the middle of feature development, and the team could proceed without reverting any commits.

We can have a configuration file that defines a bunch of toggles for features that are pending. The running application then uses these toggles to decide whether to show or hide the feature. We should avoid protecting every code path in the new feature's code with a toggle. Instead, we can focus on the entry points that would lead users there and toggle those entry points. If we find that creating, maintaining, or removing the toggles takes significant time, then that's a sign that we have too many toggles.

Most feature toggles are set at run time, but some are set at build time. The advantage of a build time toggle is that none of the new feature's code gets compiled into the released executable. Run time toggles make it easier to set up our Pipeline and to run tests with various configurations of features. It also facilitates canary releases and A/B tests, and makes it easier to rollback should a new feature malfunction in Production.

Legacy code often misses feature flags. Teams go through a cost-benefit analysis to decide whether to rewrite/refactor that code, or add feature flags as is.

Epic 21

As an engineer
I want to spin up Pipelines from a code/configuration file
So that the Pipelines are declarative, ephemeral, versioned and traceable

We should have a well-defined Pipeline, which details OSS (Open Source Software) and/or commercial tools/technologies in use. We should also have a clear definition of customizations that were made to vendorware, so that bugs can be triaged appropriately either to internal Scrum teams or to external vendors.

It is strategic to embed solutions architects from external vendor organizations within the internal Scrum teams, given that the Pipeline is a collective venture. For customizations made to vendorware, it is helpful to go through code review sessions with external vendors, much the same way as we do with the internal teams.

Epic 22

As an engineer
I want to receive automated notifications for Pipeline failures on channels that I subscribe to, both on my mobile and non-mobile devices
So that I can respond and fix within the desired SLA (Service Level Agreement)

For humans to intervene, Pipelines should generate:

- Alerts, for suspicious events

- Notifications, for failed events

While alerting and notifying, we should be mindful of fatigue caused by spam. If we inundate everyone for everything, folks end up being less responsive than they would have been if the notices were smartly targeted to potential culprits.

Epic 23

As an engineer
I would like to consider a topic-based Publish Subscribe architectural pattern for inter-system communication
So that publishers and subscribers can communicate through a Topics Broker

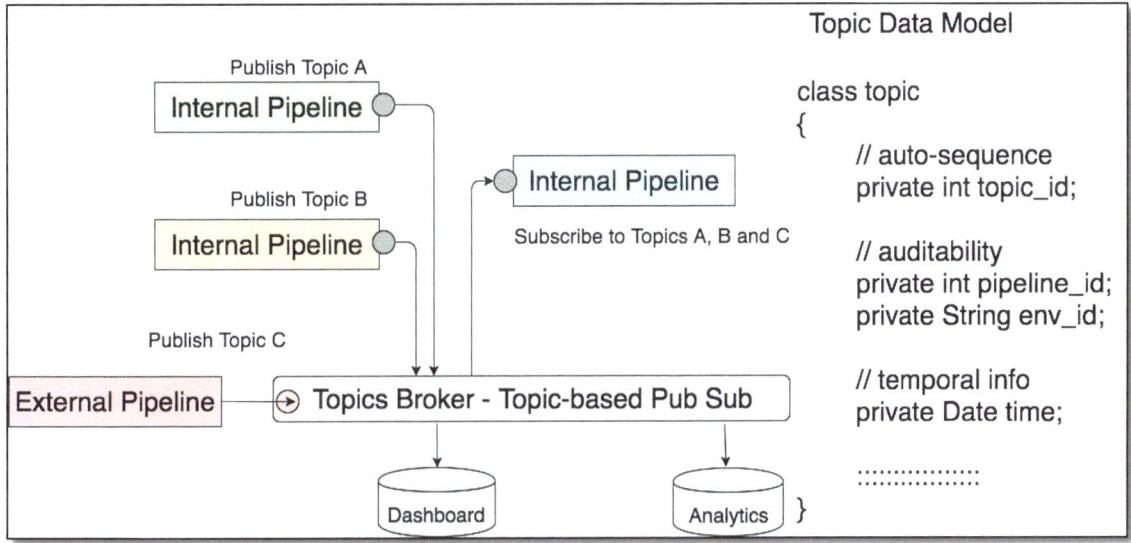

Figure 14 - Publish Subscribe Architectural Pattern

View, print and download from http://continuity.world

Publish Subscribe, or Pub Sub, can help:

- Trigger downstream Pipelines, which could be parts of different systems that appeared out of mergers and acquisitions

- Log KPIs which then provides downstream analytics and insights

- External vendors post topics to the organization's internal Topics Broker in a controlled manner to report upgrades to vendorware consumed by the organization

Epic 24

As an engineer
I want to enable a mechanism for external vendors to publish topics to the organization's Pub Sub broker
So that internal teams, whose products depend on external vendorware, can subscribe to topics that reveal when and how vendorware is updated

Epic 25

As an engineer
I want to establish a data model for topics (if a topic-based Pub Sub is used)
So that published topics can be interpreted by interested subscribers and audited for security

Epic 26

As an engineer
I want to establish a data model for Subsystem (the smallest deployable and runnable units) Manifests
So that there's transparency over the:

- Loosely coupled components (the smallest distributable and testable units) that make up each subsystem

- Build number that led to a certified version

- Environment that was instrumental in execution

- Owner who can be contacted, if needed

On a similar note, a System Manifest can be used to define a system assembled from loosely coupled subsystems.

Epic 27

As an engineer
I want my Pipeline to perform ZDD (Zero Downtime Deployment) on Dev, Stage and Production environments
So that both internal and external users are least affected when new product releases are done

Blue-green deployments are a specific kind of ZDD (Zero Downtime Deployment), where

- All users are currently pointing to blue with version n at time t1

- A production deployment is made to green with version (n+1) at time t2>t1

- 0.01% traffic is routed to green, so that in the event of problems, only a minuscule percentage of users will get affected

- Automated validation is performed on green

- If the validation is successful, we gradually switch 100% traffic to green in increments of 5 to 10% and continue to run automated validations

- If the validation is unsuccessful at any time, we revert the 0.01% traffic (or however much traffic has been routed thus far) back to blue, thereby causing minimal disruption to customers

Epic 28

As an engineer
I want to have configurable verbosity (Info | Warning | Error) in the Pipeline logs
So that I can troubleshoot with ease and respond/fix within the desired SLA (Service Level Agreement)

Epic 29

As an engineer
I want the Pipeline to leverage the circuit-breaker pattern
So that when an external system is unavailable, the Pipeline can recover from communication failures gracefully

Some standard use cases where Pipelines interact with external systems are:

- A source control repository like Atlassian's Bitbucket or GitHub

- A dashboard, like SumoLogic or ELK Stack, to log KPIs (Key Performance Indicators) that enables the leadership team to make informed decisions

- An artifact repository to publish versioned artifacts, like Sonatype's Nexus Repository or JFrog's Artifactory or Amazon's S3

- On-prem data center VMWare VMs or Amazon's AWS or Microsoft Azure or Pivotal's Cloud Foundry Platform as a service (PaaS) or Heroku PaaS to deploy and execute tests

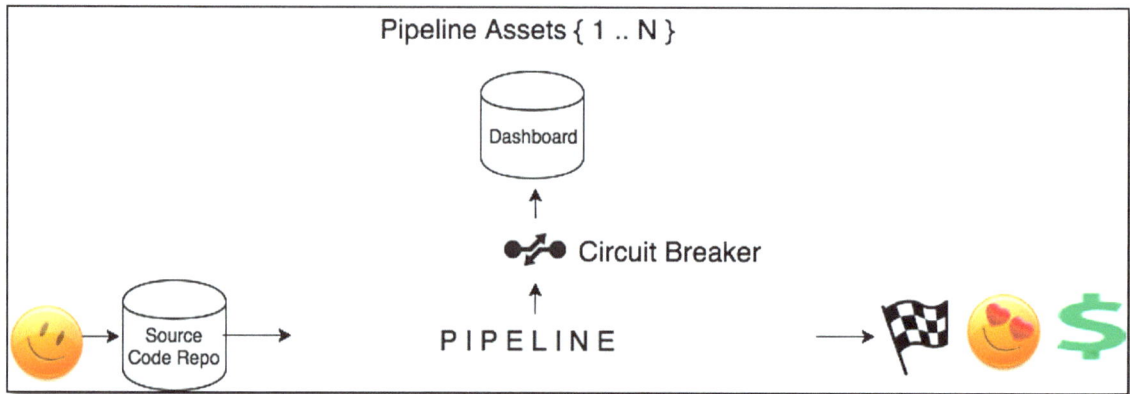

Figure 15 - Circuit-breaker Pattern

View, print and download from http://continuity.world

- Atlassian's Jira to manage bugs and change requests

- A configuration server to manage configuration

- A web-browser and mobile device farm, like SauceLabs or Amazon Device Farm, to execute web and mobile functional tests

- An external SaaS service like BlazeMeter to run performance tests

- Etc.

Epic 30

As an engineer
I want to define and monitor KPIs in the Pipeline infrastructure
So that these KPIs can be trended over time and informed decisions can be made

An example Pipeline KPI could be Check-In-2-Go-Live. Check-In-2-Go-Live is a rough summation of Check-in-2-Dev, Dev-2-Stage and Stage-2-Production, along with the wait at the manual gate(s), if we have any. Let's assume that there are three environments - Dev, Stage and Production - and we could apply this same principle to all environments that we have.

For more details, see chapter "Continuous Analytics And Insights".

Epic 31

As an engineer
I would like to design an optimized datacenter topology
So that network latency is minimized when the Pipeline executes from Dev to Stage to Production and interacts with internal systems and external SaaS/IaaS/PaaS providers

For Product Owners

Epic 1

As a Product Owner
I want to define the budget for the Continuous Delivery Pipeline product and project
So that construction and maintenance work can be paid for

Epic 2

As a Product Owner
I want to prioritize epics that are related to tech debt, digital transformation, DevSecOps and Continuous Delivery
So that culture and processes improve with technological advances

Epic 3

As a Product Owner
I want to see a Demo for the Pipeline as per the DoD (Definition of Done) established below
So that I understand:

- The product architecture of the slice we have chosen to continuously deliver

- The test architecture of the product slice we have chosen to continuously test

- List of tools that are home-grown, OSS (Open Source Software), * as a service

- Maintainers of tools and domain services used to construct the Pipeline

The following use cases demonstrated via a Pipeline visualizer reflect an operational Pipeline:

- Team starts with a fully flowing Pipeline

- Team deliberately checks in erroneous product code, along with a good test, and chokes the Pipeline in Dev

- Team checks in the fix and the Pipeline flows again

- Team deliberately checks in defective new feature code, with no legitimate test, and chokes Production

- Team demonstrates automated rollback in Production

- Team deliberately checks in defective test code/config/data and chokes the Pipeline (A test bug should not be confused with a product bug)

- Team checks in the fix to the test and the Pipeline flows again

- Team deliberately checks in defective infrastructure code and chokes the Pipeline (A foundation bug should not be confused with a product or a test bug)

- Team checks in the fix to the infrastructure and the Pipeline flows again

- Team reviews code/configuration for a Pipeline definition that spins up the Pipeline automatically

- Team demonstrates Pipeline KPIs, like Check-In-2-Go-Live, being logged to a dashboard

KEY TAKEAWAYS FROM CHAPTER 5

- Everyone in the organization has a part to play in Continuous Delivery, and now "Continuous Everyone" is a thing!

- Avoid maintaining a separate Continuous Delivery Pipeline backlog. There should be one single prioritized backlog for the whole organization, or else execution will be fragmented and frictional.

- The DoD (Definition of Done) is the key to success. Half-baked DoDs lead to budgets running out, premature celebrations and Pipelines that choke.

Chapter 6 | The Twelve-Factor Pipeline

The Pipeline is an incarnation of Process-As-Code.

It is a product in its own right.

It is a first-class citizen, or else why should we release quality products to our customers through it?

Sometimes we fail to do due diligence in treating the Pipeline as an application, since it is not our core domain. However, Pipelines reflect hygiene (or the lack of it) and impact culture, productivity and velocity of our teams.

Inspired by The Twelve-Factor App (https://12factor.net/), the Twelve-Factor Pipeline establishes the same gold standards for the Pipeline application, as are applied to the products that flow through the Pipeline.

i. Codebase

One codebase tracked in revision control, many deploys

Codebase for the Pipeline should be checked into our source code repository, much like codebase for the rest of our applications.

Pipeline codebase could include code, scripts and configuration related to the Pipeline's:

- DSL (Domain Specific Language)

- Domain services

- OSS (Open Source Software) components, along with customizations

- SaaS/IaaS/PaaS components, along with customizations

Pipeline DSL could be YAML, JSON, XML, Groovy and other declarative formats that are advocated by the Pipeline providers.

Depending on the exact nature of these services, domain services can be implemented as web services, REST services, AWS Lambda functions stitched together with AWS Step Functions, static polyglot functions using the Services Architecture or whatever makes sense for your organization. See "Domain Services" in chapter "Continuous Delivery As A Domain".

The focus should be on defining sharp interfaces and RBA (Role Based Access), such that segregation/separation of duties can be implemented with software gates. RBA (Role Based Access) enables organizations to institute:

- Who can make changes, and

- What part of the Pipeline constructs they are allowed to change.

For example, a junior developer inadvertently modified gating criteria to exclude performance test results within the Pipeline and that could have led to Production issues.

So:

- We should adopt RBA (Role Based Access) to establish a select group who can modify software gates, such that the rest of the organization can stay safe.

- While empowering teams to innovate at speed, it is best to institute a culture of pull requests and code reviews before merging to the Master branch of the Pipeline.

Both these measures make doing the wrong thing difficult and the right thing easy.

OSS components are integral parts of the Pipeline, and these could range from orchestrators, dependency management tools, test frameworks, code coverage measurement tools, static code analyzers, database change management tools, and the like.

Commercial vendors provide SaaS (Software as a service), IaaS (Infrastructure as a service) and PaaS (Platform as a service), and have dominated the DevSecOps space in the last decade. They help startups and enterprises, which have embarked on this journey, to focus on their products and customers instead of worrying about the logistics.

An example of an age-old malpractice is to have scripts injected into text boxes inside Jenkins freestyle jobs. The freestyle job configuration, if untraceable, can lead to unpredictable behavior. It is also unacceptable to have Pipeline scripts reside locally on servers that are not built and packaged through version control.

ii. Dependencies

Explicitly declare and isolate dependencies

Pipelines can have dependencies on:

- Cloud providers like, Amazon Web Services (AWS) and Microsoft Azure

- Jenkins Pipeline-As-Code, TravisCI, GoCD, AWS CodePipeline, Spinnaker, ConcourseCI, Bitbucket Pipelines etc. which provide an orchestration layer and the DSL (Domain Specific Language) to build Pipelines

- Docker and its ecosystem for containerization

- Kubernetes or Mesosphere or Amazon's ECS (EC2 Container Services) for container management or orchestration

- SauceLabs, which provides a farm of VMs to run Selenium-based Web functional tests and a farm of mobile devices to run Appium-based iOS and Android functional tests.

- JMeter for running performance tests

- BlazeMeter for running performance tests in the Cloud, essentially by re-using the same JMX scripts that are used for JMeter

- Hashicorp's Packer and Vagrant to help build immutable infrastructure

- Java-based test frameworks like TestNG and JUnit

- Grunt, Bower, Jasmine to build NodeJS applications

- Cobertura and JaCoCo to measure unit test coverage

- Liquibase, Datical (enterprise Liquibase), Flyway and DB Maestro to do automated change management and Continuous Delivery for databases

- Coverity to do SAST (Static Analysis Security Testing), which is static code analysis to detect security vulnerabilities in applications

- OWASP (Open Web Application Security Project) ZAP to perform DAST (Dynamic Application Security Testing)

- Cloud Foundry or Heroku, which are PaaS providers

- Hadoop, Hive, HBase, Spark, Oozie, Nifi, Apex and the ecosystem around distributed processing of batch and real-time data

- Etc.

The DevSecOps space is evolving fast and is deeply fragmented due to multiple players in the market. Tools have gone through revolving doors, causing churn in organizations adopting Continuous Delivery. However, tools have improved over the last decade or so in terms of ease-of-use and maintainability. For more details, refer to sections "NodeJS Pipeline Model", "Java Pipeline Model", "iOS Pipeline Model" and "Android Pipeline Model" in chapter "Continuous Delivery As A Domain".

Depending on the exact nature of these services, domain services can be implemented as web services, REST services, AWS Lambda functions stitched together with AWS

Step Functions, static polyglot functions or whatever makes sense for your organization. Pipelines, built as per the Services Architecture, have polyglot domain services that have dependencies on multiple programming languages. See the section on "Domain Services" in chapter "Continuous Delivery As A Domain".

Ephemeral pipelines should ideally run in containers that are spun up from predictable images and hence are traceable and auditable. Pipeline dependencies need to be explicitly declared and pinned to specific versions, such that Pipelines generate reproducible artifacts.

iii. Configuration

Store config in the environment

Sensitive configuration, like secrets, certificates and credentials, should be protected using Credential Stores and KMS (Key Management Services).

A Pipeline's sensitive configuration could involve access keys for external systems, like:

- A source control repository like Atlassian's Bitbucket or GitHub

- A dashboard, like SumoLogic or ELK Stack to log KPIs (Key Performance Indicators) that enable the leadership team to make informed decisions

- An artifact repository to publish versioned artifacts, like Sonatype's Nexus Repository or JFrog's Artifactory or Amazon's S3

- On-prem data center VMWare VMs or Amazon's AWS or Microsoft Azure or Pivotal's Cloud Foundry Platform as a service (PaaS) or Heroku PaaS or something similar to deploy and execute tests

- Atlassian's Jira to manage bugs and change requests

- A configuration server to manage configuration

- A web-browser and mobile device farm, like SauceLabs or Amazon Device Farm, to execute web and mobile functional tests

- An external SaaS service like BlazeMeter to run performance tests

- Etc.

RBA (Role Based Access) to sensitive configuration enables organizations to implement segregation/separation of duties effectively, while delivering products to customers through Continuous Delivery Pipelines. This sensitive information should include configuration for software gates that determine the gating criteria on whether code should be promoted from Dev to Stage to Production. Find more details in chapter "Segregation/Separation Of Duties".

iv. Backing Services

Treat backing services as attached resources

Pipelines should be stateless and hence do not need backing services. However, the products flowing through the Pipelines do, as per that product's architectural needs.

Pipelines should be well versed with these arrangements for backing services for the products flowing though them:

Component Phase: Design unit tests to avoid backing stores. Unit tests and Static Code Analysis need to be completed in this phase while we are the closest to the source code.

Subsystem Phase: In the Subsystem Phase, functional, performance and security tests should be designed such that in-memory backing stores are in play.

System Phase: For integration tests, performance tests and DAST (Dynamic Analysis Security Testing), Pipelines reach out to real backing stores to populate them with test data and execute tests. Access keys to these backing stores should be treated like any other secret, and should be controlled via RBA (Role Based Access). Note that we should avoid the System Phase, highly coupled product architecture and highly integrated E2E environments.

For more details on these Pipeline phases, see "Domain Model For Continuous Delivery Pipeline" in chapter "Continuous Delivery As A Domain".

Backing stores should be managed via Pipelines, as opposed to silo-ed DBA (Database Administrators) teams indulging in manual operations. Databases are subsystems (the smallest deployable and runnable units), and should be treated like applications. They comprise components or modules (the smallest distributable and testable units). They lend themselves to automated change management and Continuous Delivery with the help of emerging technologies like Liquibase, Datical (enterprise Liquibase), Flyway DB, DB Maestro etc.

Albeit enterprises have built Pipelines successfully with both SQL and NoSQL, Continuous Delivery is easier to implement for NoSQL technologies than it is with RDBMS.

v. Build, Release, Run

Strictly separate build and run stages

Pipelines are products. The Pipeline codebase should be built into versioned and traceable artifacts, just like the product that flows through the Pipeline.

Pipelines are first-class citizens and should undergo checks and balances before they can be trusted to release quality products frequently and predictably to our customers.

Pipelines are production assets and should have SLAs and 24/7 monitoring. Engineering teams are the major customers of Pipeline-As-Code and keeping them productive leads to happy and repeat customers for the organization.

So, like any other product, Pipelines need to be developed, tested and released through ... ironically, Pipelines! The Pipelines for Pipelines will have Component, Subsystem and System Phases, and software gates. Essentially, the gold standards of Continuous Integration and Continuous Delivery apply to Pipelines, just as much as they apply to other products flowing through the Pipelines. For more details, see section "Continuous Integration" in chapter "Maintaining Pipeline Domain Model Integrity".

For auditability, it will be useful to track pairs, like:

- Version 6 of the product was shipped by version 9 of the Pipeline, or
- Version 10 of the product was shipped by version 15 of the Pipeline, and so on.

{6, 9} and {10, 15} make auditable pairs for the products flowing through the Pipeline and the Pipelines themselves.

These auditable pairs make troubleshooting easier during Production issues. Let's say, version 6 of the product failed in Production due to performance issues. From the auditable pairs we see that it was certified by version 9 of the Pipeline. We can troubleshoot whether version 9 of the Pipeline executed performance tests and whether they passed. If they did pass, we would check whether the current performance suite includes this performance scenario or whether we need to introduce it, or whether the performance benchmark was inaccurate. If we do need to add/alter the tests, then we build, release and run the enhanced version of the performance test suite along with the Pipeline, and ensure that the Production issue is addressed. In this scenario, unless we changed Pipeline codebase, the version of the Pipeline would remain at 9. The version of the performance test suite will get bumped up, and the Pipeline would pick up the latest suite in its next run.

vi. Processes

Execute the app as one or more stateless processes

Pipelines should be stateless. Stateful Pipelines are brittle, and severely impede scalability.

Pipelines are instances of Process-As-Code and should be triggered on commits. Each and every commit triggering a separate Pipeline could be overkill in large teams. However, a large batch of commits that trigger one single Pipeline can hinder troubleshooting in the event of failures. We need to find the optimum window for our teams.

Depending on this batch window and the rate of commits of the team, a variable number of Pipelines can execute at any given time, and hence auto-scaling the Pipeline cluster can help. This is the evergreen CAPEX vs. OPEX conversation that no longer is hard to validate with the emergence of Cloud technologies like AWS (Amazon Web Services).

Pipelines should be ephemeral and should spin up in containers:

- As and when needed, and

- On any node as needed - be it on our laptops or in the Cloud or on VMs inside a Data Center

This is possible only if there's no state.

However, there are three things we need to keep in mind, as far as statelessness is concerned:

#1 – Preserving Pipeline Execution History And Audit Trail

Statelessness should not be confused with the Pipeline's ability to persist test and audit evidence in a System of Record. For example, we should be able to view test statistics, build statistics and KPIs for every Pipeline that executed:

- In the last X days, or
- The last Y times.

X and Y are configurable figments of the retention policy that we choose for the organization based on storage constraints. Storage has become inexpensive and hence isn't much of a constraint. However, be mindful since things add up over Sprints, months, quarters and years.

#2 – Avoiding Cumbersome Downloads Of Dependencies Per Commit

Statelessness should also not be confused with the Pipeline's ability to "store" downloaded artifacts between runs. Java shops typically have dependencies on libraries that are cumbersomely downloaded when the Pipeline builds versioned artifacts, and this can be overkill for every commit-based Pipeline. This can be handled by taking advantage of a local file-system. And this same requirement is applicable for containerized Pipelines. Albeit it's a known fact that containers do not have persistence built in, a mounted volume can give the notion of persistence inside a container.

The caveat is that it is more predictable for a Pipeline to be a clean slate. We need to make a conscious trade-off between having a clean slate and lowering the build execution time.

#3 – Marking Off False Positives Reported By Static Analyzers

Another case where Pipelines need to "remember" is when static analyzers report false positives and they need to be marked off so that subsequent builds do not report them. A layer of persistence should be added that would record all static code analysis issues, along with flags to indicate false positives. Metadata around who marked them off, on what date and an explanation of why it is harmless should also be available for audits.

vii. Port Binding

Export services via port binding

The Twelve-Factor Pipeline should follow guidelines articulated at the Twelve-Factor App's "Port Binding" section: https://12factor.net/port-binding.

viii. Concurrency

Scale out via the process model

Pipelines should execute on a commit-based model, whereby a Pipeline is triggered when an engineer checks in code, configuration, tests and/or data.

Every single commit triggering a separate Pipeline could be overkill for large teams and a time window should be defined within which commits would be batched up. This is in violation of the ideal state where every commit makes it to Production. However, we need to be practical and not all enterprises have ideal circumstances. Commits are batched during this "quiet window", as it is often called. The "quiet window" shouldn't be too large such that granularity of commits is lost and troubleshooting takes longer.

The rate of commits per individual, team or organization is variable. Hence, we do not know the number of Pipelines that need to be spun up at any point in time. We can look at historical data; however, past performance may not be an accurate indicator of the future.

Pipelines should be ephemeral and should spin up in containers:

- As and when needed, and

- On any node as needed - be it on our laptops or in the Cloud or on VMs in the Data Center.

This is possible when Pipelines avoid keeping state between runs. Stateless Pipelines help scalability too. Auto-scaling features of Cloud technologies can ease the need to budget capacity for an unknown concurrency.

CAPEX (Capital Expenditure) leads to upfront budgeting, whereby we can set up a cluster of nodes that offer distributed and parallel processing. The capacity of each node and the cluster are calculated based off past trends and could be re-assessed at a later time. At a subsequent budgeting session, we could re-evaluate capacity and addition/subtraction could be made based off actual data.

OPEX (Operating Expenditure) can entertain scaling out, as and when needed. Most organizations prefer the OPEX model due to financial constraints and tax liabilities. They pay for resources that were actually used and hence see variable bills based off fluctuating demand and supply.

There's an important side-note to all of this with regard to concurrency. When we say that we prefer to have Pipelines run concurrently, we mean Pipelines allocated to different artifacts. Typically the mapping between an independently deployable artifact and a Pipeline is 1:1. Let's say we have N artifacts, and N Pipelines that map to each of those artifacts. All N Pipelines could and should run in parallel, if needed, since all the artifacts could be upgrading at the same time. However for the same artifact, let's say artifact 1, several instances of Pipeline 1 may not be able to execute at the same time. For example, when instance 1 of Pipeline 1 is making a deployment to a static instance of EC2, instance 2 of Pipeline 1 should wait or else the deployments would step on each other, leading to unpredictable results. Pipeline orchestrators have built-in queuing mechanism to prevent such race conditions.

ix. Disposability

Maximize robustness with fast startup and graceful shutdown

Treat Pipelines like cattle, not pets.

I am not particularly fond of how animal life is being referred to here, however, the underlying thought is clear. Ephemeral Pipelines should be easy enough to spin up and inexpensive enough to discard. Malfunctioning Pipelines don't need to be nursed back to health and teams can spin up new ones.

Typically, Pipelines are triggered:

- On commits (or, granular batches of commits) to the source code repository

- On schedule, if business cases so demand.

- Manually, if the situation so demands. Ideally, automation is the way to go.

A small caveat though. The overhead of spinning up Pipelines adds up over time, and care should be taken to optimize the total execution time. However, optimization problems are better handled later in the game, and only by the best minds. It is important to prove out the functionality first, so that our teams start reaping RoI (Return on Investment) as soon as possible.

In the event that a Pipeline fails to execute, it could be programmed to retry. However, if there is a legitimate test failure, there is no point in automated retries until someone fixes the bug.

Similarly, if the Pipeline is unable to complete execution within the stipulated threshold based off a configurable timeout parameter, it could be programmed to retry. If timeouts happen often, we could consider adjusting the timeout value after carefully diagnosing the cause of delay. However, without appropriate reason, do not blow up the total execution time by inflating timeouts.

After a pre-meditated number of retries, Pipelines should terminate on their own, or could be terminated by the team manually. Like any other application, care should be taken to release all unused resources.

x. Dev/Prod Parity

Keep development, staging, and production as similar as possible

The Pipeline should be environment-agnostic. Ephemeral Pipelines, running in containers, should behave the same way on our laptops as in the Cloud, or even on VMs in the Data Center.

This can be achieved if we:

- Manage secrets, certificates and credentials, and for that matter all sensitive information, as configuration external to the Pipeline code

- Keep the Pipelines stateless

Pipelines should behave in an identical fashion on Dev, Stage and Production, so that we can steer clear of classic excuses like "works for me". Moreover, the cost of a bug is low in Dev, medium on Stage and high in Production, and hence we should "find fast, fix fast".

The ability to validate Pipelines locally improves the team's productivity since fixing mistakes in isolation is inexpensive compared to when it becomes everyone else's problem.

Also, the probability that a bug gets branded as a "feature" increases towards the later parts of the Pipeline. So, we should "catch 'em young!". And the only way we will catch all the bugs on Dev is when Dev is identical to Stage and Production.

xi. Logs

Treat logs as event streams

Pipeline logs expedite troubleshooting, enable monitoring and facilitate audits.

A high verbosity is better than low when we are starting our journey of Continuous Delivery. A talkative Pipeline is more likely to reduce/eliminate time spent in reproducing failures than the silent ones. However, once our processes have matured, we can reduce the verbosity, since Pipelines are commit-based and can quickly aggravate the I/O overhead and storage cost.

Pipelines should log KPIs (Key Performance Indicators) to dashboards like ELK Stack and SumoLogic, once we perform the relevant instrumentation. A typical Pipeline KPI is "Check-In-2-Go-Live", which reflects how long it takes for a check-in to go live from Dev through Stage to Production. Check-In-2-Go-Live when trended over a week, month, quarter or year, can be a crude reflection of velocity and hence Time2Market. FeatureLeadTime is a better metric to trend velocity though. See the section on "Concept-2-Cash" for more details in chapter "Continuous Analytics And Insights".

Pipeline logs can also be mined for insights with tools like Splunk. These insights could be sent as notifications to the right audience in a proactive manner, thus preventing potential failures in the future.

xii. Admin Processes

Run admin/management tasks as one-off processes

Admin or management tasks are critical for Pipeline maintenance, audits and analytics. Refer to chapter "Continuous Analytics And Insights" for details on KPIs (Key Performance Indicators) that benefit organizations.

Examples of Pipeline admin/management tasks are:

- Calculating repeat offenders – both people and tasks. For example, engineers with a track record of breaking greater than 20% of Pipelines triggered by their commits in the last month could be reported as repeat offenders and a training session could be organized to coach them on how to perform more validations in their local environments. Or, a deployment step that has a failure rate greater than 20% could be analyzed to identify the root cause, since some deployments fail due to configuration management issues or environment issues, which then should be tracked closely.

- Calculating slowcoaches. Pipelines are meant to provide timely feedback. Pipelines that take longer than X hours to finish should be reported, so that someone could work on optimizing the execution time. X depends on your organizational goal.

- Periodic archiving of test or audit evidence in a System of Record.

Code related to admin/management of Pipelines should be organized cleanly in the Pipeline codebase, and should be versioned, tested and released along with the Pipeline application itself.

KEY TAKEAWAYS FROM CHAPTER 6

- Hear no manual, see no manual, and speak no manual. The Pipeline is Process-As-Code.

- The Pipeline is a product in its own right and is designed using the Services Architecture.

- The Pipeline is a first-class citizen and the gold standards of software development apply to it, just as much as they apply to the products flowing through the Pipeline.

CHAPTER 7 | SEGREGATION/SEPARATION OF DUTIES

Segregation or separation of duties or powers is a controversial topic in organizations that have dwelled on a legacy hierarchy long enough to confuse separation of duties with separation of departments. These organizations experience severe delays in implementing Continuous Delivery Pipelines, since Pipelines tend to relentlessly gnaw through departments and silos, and rightfully so, and can be seen as posing risk to the business and our customers.

This chapter explains the "what" and "why" of Segregation of Duties and why it is *critical for governance*. Continuous Delivery Pipelines improve speed, quality and predictability of our product releases and are *critical for sustainability*. So, trading one for the other is not an option.

Hence, we will model Pipelines to honor Segregation of Duties (SoD) and deliver quality products frequently and predictably without risking business.

SoD In The US Government

To understand segregation/separation of duties, let's take a close look at how the United States government operates. The government's power is separated into three branches to prevent one person or group from having too much control.

The three branches of the government are:

- Legislative
- Executive
- Judicial

The separation of government into three branches creates a system of checks and balances. This means that each branch can block, or threaten to block, the actions of the other branches.

Consider the following examples of checks and balances:

- If the President, who heads the Executive branch, signs a questionable treaty, the Senate, who is part of the Legislative branch, can block it.

- If the Congress, who is part of the legislative branch, passes an unconstitutional law, the US Supreme Court, who is part of the judicial branch, can block it.

This separation of powers limits the power of the government and prevents the government from violating rights of the American people.

SoD In The Enterprise

SoD in enterprises:

- Prevents the enterprise from violating the rights of employees.

- Prevents the enterprise from violating the rights of customers.

- Is needed for unbiased governance, where one person or group does not become too powerful.

SoD has been problematic in the world of Continuous Delivery and Deployment, and is a hindrance for DevSecOps evangelists trying to design and implement smart Pipelines. The evangelists understand the need for SoD; however, they do not endorse the interpretation that Continuous Delivery poses a threat to their business and customers.

The bottom line is that we need both SoD and CD in enterprises. So, we will tease the problem apart and model SoD in the Continuous Delivery Pipeline domain, such that we will have responsible speed and not suicidal speed.

Current State Of DevSecOps And CD

Let's review what happened in the last decade or so in enterprises that are in the process of incorporating DevSecOps principles. The burning question is around closing the skills gap, as in; did DevSecOps personnel exist before DevSecOps?

"DevSecOps" is a software building, testing and releasing process where product owners, development, security and operations collaborate and communicate from ideation to delivery. It is a paradigm shift to how quality software is released frequently in a predictable manner through Continuous Delivery Pipelines.

Long time ago, in a galaxy far, far away, we had Dev, QA, Tools, Infrastructure, Platform, Release, InfoSec, Operations and whatever you have, passing the baton much like in a relay race. And there were Program/Project/Release Managers "managing" the release process through meetings and emails, so that batons won't fall through the cracks.

Even today, there are some organizations that have fragmented their organization such that the Continuous Delivery Pipeline has to jump over walls to pass the baton. And when we fail to pass the baton successfully, our customers ding us, much like how the authorities disqualify the relay team.

Eventually, teams got tired of dropping the baton and started dreading the subsequent penalties. QA teams started partnering with Dev, Tools teams got (evenly or unevenly) distributed everywhere, Release teams started merging with Platform, Infrastructure teams started coalescing with Operations and so on and so forth.

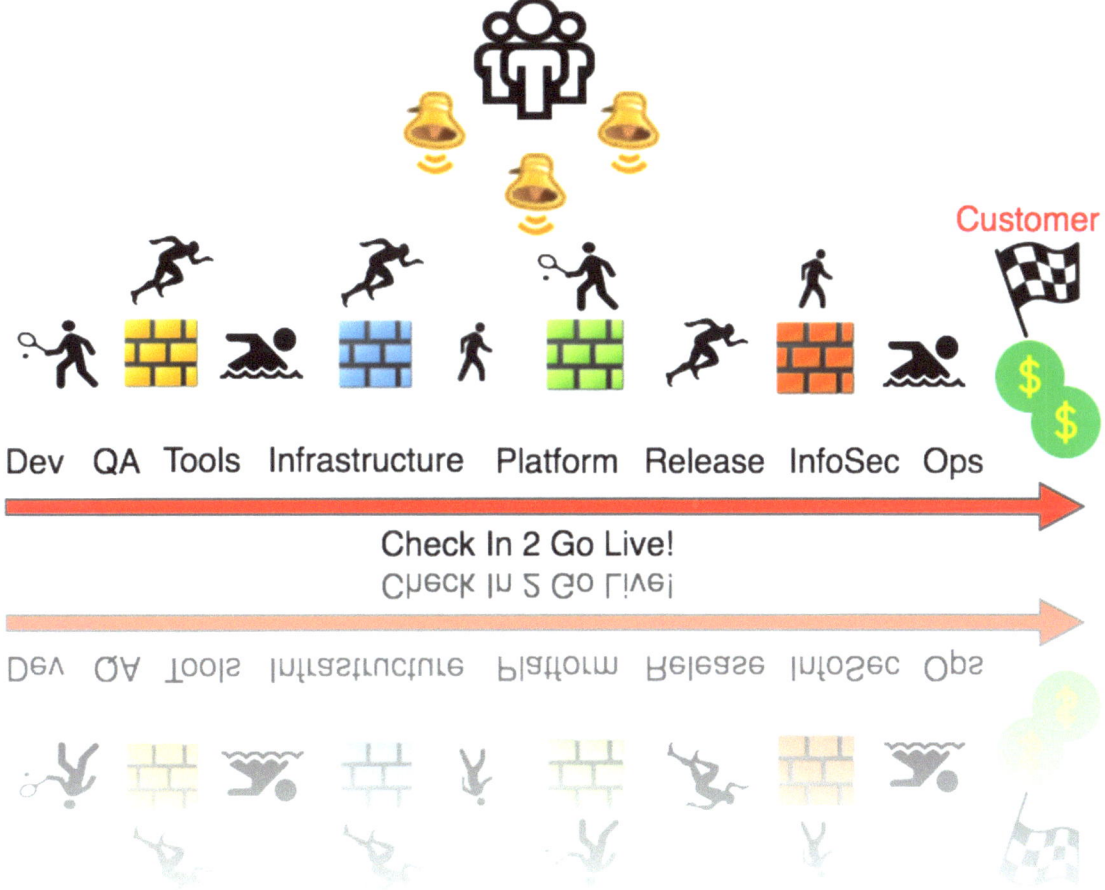

Figure 16 - Hand-off Anti-pattern

View, print and download from http://continuity.world

The exact nature of the partnerships could be different in different organizations. What is more important is that people started learning new skills and moved out of their comfort zones. Some took the *"Tour of Duty"* and moved around till they got a 360-degree view of the largely fragmented Engineering organization.

In the truest sense, DevSecOps means:

- Scrum teams own their code from Dev to Stage to Production.

- We do a "*Left Shift*" by pulling security upfront. We build security into our applications from the design phase, instead of evaluating a finished product.

- Releases happen through Continuous Delivery Pipelines in a boring and uneventful manner.

- KPIs like FeatureLeadTime and Check-In-2-Go-Live are low, and we can experiment with our ideas frequently and safely. And more importantly, we can recover from failures fast.

DevSecOps is a paradigm shift and despite our adoption attempts is still written as "DevSecOps" (within quotes), since we are largely in the process of internalizing it. One of the major hindrances in adopting Continuous Delivery has been SoD (Separation of Duties) and hence we will invest into modeling it in the Pipeline domain.

Core Principles Of SoD

Let's get to the crux of SoD, so that we can do an impartial design that addresses all outstanding concerns. When an enterprise falters on implementing SoD in CD, we should revert to these principles instead of seeking human approvals and manual sign-offs.

Segregation/separation of duties (SoD) essentially comprises four concepts, and we should design and implement Pipelines to address these so that CD and SoD can co-exist.

The four pillars of SoD are:

- The concept of having more than one person/group required to complete a task

- Separation of powers needed for unbiased governance

- An internal control intended to prevent fraud and/or error

- To disseminate tasks and associated privileges for business processes amongst multiple users

Modeling SoD In The Pipeline Domain Model

Refer to section "Domain Model For Continuous Delivery Pipeline" in chapter "Continuous Delivery As A Domain".

In organizations that have Dev/QA/Release/Operations/InfoSec/etc. headed by separate chiefs in the C-Suite (Chief Product Officer, Chief Technology Officer, Chief Information Officer, Chief Information Security Officer etc.), SoD could be interpreted as one department having to manually approve or sign-off on another department's work. Segregation of duties is very different from segregation of departments, although ironically both are represented by the acronym SoD! This leads to manual hand-offs and coordination overhead that eat away productivity and velocity.

The overarching question is: if Scrum Teams own their code all the way from Dev to Stage to Production, is this group becoming too powerful, and is this a violation of SoD?

It could be, unless we model SoD into our Continuous Delivery Pipeline domain in the following ways:

SoD Core Principle #1

The concept of requiring more than one person/group to complete a task

For those of us familiar with the Stock Market, we understand the value of building a diversified and balanced portfolio, whereby we include different securities to mitigate risk.

According to http://www.nasdaq.com/investing/glossary/w/well-diversified-portfolio, a well-diversified portfolio includes a variety of securities so that the weight of any security is small. The risk of a well-diversified portfolio closely approximates the systematic risk (http://www.nasdaq.com/investing/glossary/s/systematic-risk) of the overall market, and the unsystematic risk (http://www.nasdaq.com/investing/glossary/u/unsystematic-risk) of each security has been diversified out of the portfolio.

Along the same lines, we can build indices that include various metrics that contribute to the overall index, and at the same time, prevents any skew/bias towards any one single metric.

See "Weighted Stability Index" in chapter "Continuous Analytics And Insights" for an example of a triangular index. This index has three metrics that are typically championed by different players in the enterprise, who have high stakes in stability. If you ask for a definition of stability from key stakeholders, you are guaranteed to get different responses, and this index enables you to factor in all responses that are logical and scientific.

See "Weighted Code Quality Index" in chapter "Continuous Analytics And Insights" for an example of a hexagonal index. This index has six metrics that are championed by different players in the enterprise. This index typically factors into one of the earlier software gates in a Continuous Delivery Pipeline, and prevents "Garbage in, Garbage out" from blowing up the Pipeline.

These are samples, and your indices should fit your needs and should include metrics that are relevant to what success looks like to you. The indices could be weighted as per your business needs, however they don't need to be in case your business decides to provide all KPIs a level playing field.

SoD Core Principle #2

Separation of powers/concerns needed for unbiased governance

In the US Government, there are three branches: Judicial, Executive, and Legislative.

In enterprises, it is unclear who the powers are due to a lack of constitutional guidelines. Even if we were to come up with something like a "DevOps Constitution", there would be no way to enforce it. We understand the role of the CEO and the C-Suite at a high level. However, each enterprise defines its C-Suite as per their business needs and Continuous Delivery gets spread across the Chief Technology Office, Chief Information Officer, Chief Product Officer, Chief Information Security Officer etc. There's nothing fundamentally wrong with a cross-functional initiative having shared ownership and goals, however, to define standards for separation of powers, we need to understand:

- Who the powers are

- Where the powers apply, and

- When the powers are invoked.

It is unclear who the exact powers are, since not all enterprises have all the chiefs at all times.

However, we could tilt this debate from who the powers are to who the powers should be. We should honor *Conway's Law*, which states:

"Organizations which design systems ... are constrained to produce designs which are copies of the communication structures of these organizations".

Once we have our product architecture, we should design our organization to reflect that architecture. Essentially, architecture should drive organization, not the other way round. Additionally, our organization should be designed in a way that decisions flow efficiently. The overarching goal is that our product(s) should be delivered with speed, quality and predictability to our customers by our Scrum teams.

In addition to discussing who the powers are and who the powers should be, let's review where the powers could be. Powers, both in life and at work, are almost always at pressure points. See Figure 1 in section "Domain Model For Continuous Delivery Pipeline" in chapter "Continuous Delivery As A Domain". Note the gates where code gets promoted from one phase to the other, for example, from Component Phase to Subsystem Phase, or eventually to the Production Phase. Those gates are the pressure points, since that's where code promotion decisions are made. Also, see Figure 3 in section "Pipeline Domain Model Fragmentation" in chapter "Continuous Delivery As A Domain" to see examples of software gates, which open only when certain checks and balances have gone through. The checks and balances could be some combination of Unit Tests, Static Code Analysis, Functional, Integration and/or Security tests, depending upon where the software gate is within the Pipeline.

An important thing to remember, while separating concerns, is that DevSecOps advocates "*Left Shift*", which implies we should build the toughest gates first. From Pre-Commit to Dev to Stage to Production, we should build quality and security into the product design, such that defects are identified and fixed by Scrum teams early in the game. We should move away from the existing malpractice where a finished product is submitted for certification, whereby bugs get promoted to features.

SoD Core Principle #3

An internal control intended to prevent fraud and/or error

Software gates are internal controls designed to open/close based on metrics. A Continuous Delivery Pipeline interfaces with software gates to decide whether to promote software from one phase to the next.

Let's review a concrete example, where a Pipeline interfaces with the main gate, which integrates with four child gates.

Main gate = Function of (child gate 1, child gate 2, child gate 3, child gate 4), where

- Child gate 1 validates performance

- Child gate 2 validates application security

- Child gate 3 validates functional use cases, and

- Child gate 4 validates integration interfaces and the network.

These four child gates could have different stakeholders in the organization, and we could have more child gates (or less), depending on our business needs.

Note that the gates could be binary, and could simply return the integer:

- Zero (0) for success in executing all tests

- One (1) for failure in executing at least one test

However, some teams like to know how many failures there were overall, or the type of failure, for example, functional or performance. Hence the failure return codes could be more sophisticated than being just the integer one. The child gates could each return zero for success, or non-zero for failures to the main gate. The main gate could sum up the return codes of the child gates, and hence could return zero for success, and non-zero for failures to the Pipeline. When the Pipeline receives a zero success code from the main gate, it promotes code from that phase to the next, or else it aborts with proper alerts and notifications.

For complex requirements, you could consider different weights for different gates, and a function more sophisticated than summation. However, try to keep the design simple. For details on metrics, indices and KPIs that contribute into the making of impartial software gates, refer to Figure 11 in chapter "Continuous Analytics And Insights".

SoD Core Principle #4

To disseminate tasks and associated privileges for business processes amongst multiple users

For details on business processes, refer to Figure 2 in section "Domain Services" in chapter "Continuous Delivery As A Domain". In this same context, it is crucial to understand the role of ACL in section "Anticorruption Layer" in chapter "Pipeline Domain Model Integrity". In both these figures, we see the tasks/services that need to be disseminated amongst multiple users. While we review the business processes, note that everyone is involved, whether they know it or not, thus underscoring the concept of *"Continuous Everyone"*.

Associated privileges could be translated into credentials, secrets, certificates, sensitive configuration and the like that control access to external systems that interface with the Pipeline. A common concern that pops up is the handling of deployment constraints via automation. For example, a retail giant typically freezes Production deployments during certain weeks of the year when consumer spending is predicted to skyrocket. That is the reason (excuse?) they use to mandate human approval right before every Production Deployment. This "problem", if we may deem it so, can be addressed by the ubiquitous "calendar" utility that can be configured to recognize Production blackout dates. The privilege to configure the calendar could be restricted with RBA (Role Based Access) to a select group of company personnel; however, all Pipelines could integrate with this "Deployment Calendar" such that automated queries could validate whether the system date is a blackout date. If not, Production deployments could proceed as usual. If it is, Pipelines could queue up and wait, or could abort, as per the business requirements.

KEY TAKEAWAYS FROM CHAPTER 7

- Pipeline-As-Code enables the design and implementation of segregation of duties or separation of concerns.

- Avoid human gates and human opinions. Gate yourself with software and KPIs instead.

- Organizational facts should be driven by KPIs. Avoid departmental opinions.

- Secrets, certificates, data (PII, Financial), configuration, logs and software gates should be protected by RBA (Role Based Access).

- Avoid seeking or giving approvals. Automate your resilient processes, trust them to do the job, and verify that they did.

- Audit trails should always be ON. Always. There is no reason to turn them OFF at any time under any circumstances. Period.

- Honor Conway's Law. Architecture decides organization. Not the other way round.

CHAPTER 8 | MYTH BUSTERS

Some organizations fall prey to age-old beliefs that have inherent flaws. Old-school technologists delay ushering in the new by championing outdated processes and techniques. This results in modernized approaches getting the boot and next generation technical leads stumbling over bureaucracy and red tape.

This chapter pulls up the most damaging myths and addresses the elephant(s) in the room. Education is key, and we should discuss painful issues in an open forum till the light bulbs go on. We should invest in having one-on-ones with traditional and orthodox technologists, and find a happy medium to enable experimentation of new and daring ideas. As a last resort, if experimental ideas continue to get trampled under myths, we should take firm and disciplinary measures to get the organization back on track for a successful digital transformation.

Continuous Everything Is Overkill

Still? Let's re-think.

Software is eating the world is no longer accurate. It has already consumed the world! Whether we are in the business of hotels, taxis, automobiles, medical devices, identity, cyber security, retail, advertising, finance, food, or something else, every company at the end of the day is a technology company. For the business to grow in a sustainable fashion, we need to experiment with new ideas safely and frequently.

Think about this - we put our loved ones in an automobile when we have reasonable trust in the automobile. The same way, to ship our revenue-earning products through the Pipelines, we need to trust our Pipelines. Treat the Continuous Delivery Pipeline as a product, even if it is not your core domain, for the business to grow sustainably.

Let's continue to build on this analogy. Can we afford to put our loved ones in a first class or business class seat in an airplane every time? Similarly, can we ship all our products through world-class Pipelines every day? May be not. We may have to build an inexpensive version of a Pipeline at first, and establish RoI. Once we grow our business sufficiently, we can modernize our initial prototype to a world-class Pipeline.

When we shoot for the stars, we get to the moon. If we shoot for Continuous Delivery, we get to at least Continuous Integration. If we shoot for Continuous Integration, we end up with a few more unit tests than we started with. And that's not enough.

So, think *BIG*.

Cannot Start Due To Manual Tests

Continuous Delivery can be implemented with manual gates, whereby we allow the Pipeline to poll or wait for human response. This is unlike Continuous Deployment, where there are no manual gates. Thus, we can start Continuous Delivery even when we have manual tests.

While implementing Continuous Delivery, we automate tests of the following nature to ensure quality and security of our products:

- Unit tests

- SCA (Static Code Analysis) to detect deviations from coding best practices

- Functional tests

- Integration tests

- Performance tests

- Application Security tests

    - SAST - Static Analysis Security Testing

    - DAST - Dynamic Analysis Security Testing

    - Penetration tests

We should set the bar high and automate everything – tests, configuration, deployment, reporting, monitoring, measurement of KPIs, and other relevant functions.

Also note, tests are a moving target – we add tests for new features and (hopefully) archive tests for retired features. While we are ramping up the automation, we can institute one or two manual gates, whereby the Pipeline can poll for a human to confirm the successful execution of manual steps.

However, there are three pieces of firm advice:

1. Wrap the Pipeline's polling statement with a timeout, so that the Pipeline aborts if humans do not respond within a stipulated timeframe.

2. Put a timer on the team, so that they understand that the manual gate(s) will not be there forever.

3. Put a close watch on the duration of the poll, so that execution of the manual tests takes minutes and not hours. Teams tend to abuse manual gates and it should be amply clear that these are not backdoors.

So, the debate on whether to implement Continuous Delivery in an organization is like debating whether you need to have sustainability for your business. There is no right time. Any time is as good a time as it gets.

More Environments Mean More Quality

More environments mean more maintenance overhead for sure. More quality? Not necessarily.

Multiple static environments need the same (or similar) infrastructure to be stood up multiple times. That's an expense our business folks might frown upon, if they get into the weeds of Engineering. Customers care for the quality and predictability of the product that lives in Production and is oblivious to the number of non-Production environments it was certified on. Also, environments defined per sub-organization (Dev, QA, Release, Operations) can lead to silo-ed behavior when it comes to fixing environmental problems. As far as environments go, less the merrier.

Development should work with Operations to define immutable infrastructure, which is scalable and predictable. Infrastructure, configuration, tests and data should be treated as first-class citizens, just like code, and should be packaged and distributed as versioned artifacts and images. If we define our environments to be ephemeral, we should be able to spin up environments for Dev, QA, Sandbox, Stage, Integration, E2E, etc. and spin them down once we are done. We can keep them around for troubleshooting purposes for a limited time. We have to keep in mind that in a Continuous Delivery/Deployment Pipeline, there would be overhead in spinning environments up and down per build. If the overhead of creation and destruction exceeds a certain threshold, we can pre-create a warm pool of nodes, from where the Pipelines can borrow. Refer to factors six and nine, namely "vi. Processes" and "ix. Disposability", in chapter "The Twelve-Factor Pipeline" for relevant details.

Fewer environments translate to a higher probability of maintaining parity between all of them. If we were to have identical environments in Production and non-Production, some of the conversations surrounding test data management and performance test feasibility in non-Production would get muted. Refer to factor ten "x. Dev/Prod Parity" in chapter "The Twelve-Factor Pipeline".

Containerization and Cloud migration can help reduce Data Center maintenance cost, especially personnel cost. Investments in ephemeral environments can give a much-needed edge to our business, and multiple static environments will eventually become a thing of the past.

VSM Has No Implications On CD

CD (Continuous Delivery) has no direct relation to VSM (Value Stream Mapping). A Continuous Delivery Pipeline can be built as a product (or in other words, Pipeline-As-Code) without knowledge of VSM. However, prudent applications of VSM can be effective in the Continuous paradigm, and the results can be revealing.

A common misconception is that VSM is suitable for the manufacturing industry whereas Continuous Delivery is suitable for the software industry. We can safely say that the underlying concepts of VSM apply to software engineering, with a few tweaks. The value stream is the product development process and the non-value streams are the operational pieces. Lean (and mean) software development practices have been successfully adopted, and the literature is evolving. The shift from "removing waste" to "creating and delivering value" is key, and the important thing is to do both to extract maximum benefits.

A caveat is that we should not invest too much time into etching out a VSM. Drawing the VSM is a non value-adding step. The faster we can start analyzing the results and incorporating the feedback into our Continuous Delivery Pipelines, the better for the organization.

Value stream mapping analyzes both material (artifact) and information flow. "*Artifact analysis*" involves analysis of software artifacts like requirements, use cases, and change requests, whereas "*Information flow analysis*" involves analysis of information flows in the development process.

Pipeline visualization demonstrates both artifact and information flow. The Pipeline visualizer:

- Brings to life the lifecycle of artifacts flowing through the Pipeline from source code repository to Production.

- Articulates the information flowing through various phases of the Pipeline, through software gates. See more on software gates in chapter "Segregation/Separation Of Duties".

The two metrics that are associated with VSM are value-adding times and non-value-adding (or, waste) times. Continuous Delivery aims to improve the value adding times and to reduce the non-value-adding times.

Many organizations tend to automate their current process just the way it is, which leads to a significant improvement in their productivity and velocity. However, we need to be mindful of:

- The non-value-adding operational pieces that need to accompany the value-adds, and hence need to be automated.

- The waste, which when weeded out, gives the biggest bang for the buck from our Continuous Delivery Pipelines.

So, the bottom line is that a study of VSM is likely to help build resilient processes that power our business.

Teams Must Be Co-located

This is one of the many agile principles that got blown out of proportions.

Agile rests on the principles of collaboration and communication, and with modern technologies, teams that are geographically dispersed interact through technology. We do prefer to have folks near each other; however, it is by no means a showstopper if they can't.

In geographically distributed teams, we should:

- Do more working sessions on video calls than we do otherwise.

- Draw at every opportunity. "Pictures speak a thousand words". Color-coded and detail-oriented diagrams can save many flight and hotel bills. When we have something on canvas, we can point, analyze, critique and have healthy debates around making things better.

- Record meetings. We should give our remote folks every opportunity to go back to the bits and pieces that we dug into.

- Have a Slack (or HipChat or whatever messaging system is used) channel open to discuss, communicate and alert remote (and local) teammates.

The big picture is that a Scrum team needs folks who have relevant skills to deliver projects on time and within budget. When two minds connect, the geographical distance between them collapses. Trust is what powers a team, and geographical separation is no barrier.

Engineers Can't Write Poems!

Seriously? Nah!

Once upon a time, an erroneous commit created a ruckus on Stage by breaking application security tests in the Pipeline. In total despair, the failed commit sang the following tech version of "Jamaica Farewell", known as "Commit's Farewell": https://www.youtube.com/watch?v=ZZpbfabwpFU. I have always wondered how people sing when they are devastated. I have seen it in the movies, and this commit seems no different.

A Commit's Farewell To Engineering

Down the way, where the teams are gay (I mean happy!),
And the sun shines daily on the building top.
I took a trip on the Pipeline ship,
When I came to App (lication) Sec (urity), I made a stop.

Oh I am sad to say, I broke the Pipe (line) today,
It won't be up for many a day.
And Stage is down, and heads are turning around,
Explain how it happened in Stand-Up Town!!!

So, you get the idea. Now, let's get into the weeds some more.

High degrees of separation quarantined managers into their own islands, where they could no longer understand or apprehend execution problems that plagued their organizations. Sounds familiar? And this time, the engineers were at it again, however, on a much more serious note:

An Engineer's Perspective Towards High Degrees Of Separation

Twinkle twinkle little czar
How we wonder where you are
Up above the chain so high
The ground to you seems so shy.

Ideas are lying around, gathering dust,
In you managers, we lost trust.
Should we leave? Or, should we fry?
Does it make sense to let the ground cry?

And then of course, words of wisdom from none other than Master Yoda to the Jenkins Butler on the inconvenience of maintaining freestyle Jenkins jobs:

The Master's Words For The Jenkins Butler

Unmanageable freestyle jobs you've become,
Pipeline-As-Code you must be.
Features like declarative, versioning and ephemeral are some,
Or the dark side won't set you free.

The "Analytics/Insights" report we trust,
In the groove we must get.
Make happy our engineers you must,
Don't make them fret.

Three is a pattern. Engineers can write poems!

I can't imagine that our brains can be made simple, however BrainsMadeSimple (http://brainmadesimple.com/left-and-right-hemispheres.html) suggests that the left hemisphere contributes in writing code while the right hemisphere can write poetry. There, you see! Don't tell me I didn't tell you!

It is up to us to cultivate the habit to exercise both hemispheres of our brains. If we don't push on both fronts, one side of the brain may tend to remain dormant. And engineer or not, why would you want that?

Engineering Owns Vision, Not Execution

Continuous Delivery is run as a cross-functional digital transformation program since it touches every department within an organization. While architects and technical leaders lay out the vision, Program Managers are expected to drive execution.

An alarming number of times, we have seen execution come to a screeching halt, and possible reasons could be:

- The Program Manager hails from a non-technical background and has a difficult time understanding and executing on a technical vision.

- Program Managers do not have first-hand experience working directly with engineers and hence are not used to refreshing, unpackaged and unscripted engineering jargon. Engineers say it as they see it. They don't like to water it down on a regular basis just so that someone outside Engineering could report status to executives.

- Since Agile does not have a definitive slot for a "Program Manager", the Program Manager is implanted within the Scrum Team as a "Scrum Master". As in most cases, titles don't matter as much, and this could further aggravate the execution problems.

- Engineers are a smart bunch of folks and they figure things out themselves. They take pride in cracking hard problems every day and have low tolerance for non-technical folks who need handholding. Goes without saying that these same engineers do a wonderful job of ramping newbies up, however, training and spoon-feeding are two different things.

Here's an analogy to help us understand. I don't have a medical degree or relevant experience in the medical field, so I would be useless if I run a team of surgeons inside an operation theatre in a hospital. In my zeal to learn a new trade, I may interrupt with too many questions and slow the doctors down. During emergencies, if I insist on being included in crucial medical decisions, the patient could die. I should be mature enough to admit my deficiencies and let the team accelerate without me. However, my admission or confession could lead to bad performance reviews and/or public humiliation. So, I can empathize with Program Managers and acknowledge the slippery slope they are on.

On the other hand, engineers, who get frustrated working with non-tech folks, face rebuttal for being less inclusive in their work habits. They would rather rotate the Scrum

Master role between themselves and move fast, instead of being tied down in status meetings.

In situations like these, brace for impact, since low morale leads to poor productivity. We could also lose out on customer satisfaction and employee engagement.

In summary, vision without execution is fruitless. Customized frameworks where Scrum teams can own both vision and execution are necessary for the business to succeed. When there is business value, it is important to re-imagine titles. For this reason, we have seen mature Product Owners, Tech Leads and engineers double up as the Scrum Master to enable smart execution and enhance customer delight.

*aaS Is Difficult To Embrace

Infrastructure/platform/tools vendors provide what we have to otherwise design, build, test, measure, monitor, release and maintain. However, some of us choose to believe that our products are special and external solutions will not be able to do justice to them. This leads us to write way too many homegrown solutions that inflate the TCO (Total Cost of Ownership), when adopting *aaS solutions could have been more efficient.

Here are some issues which hinder adoption of *aaS solutions:

- The *aaS solution could potentially replace someone's brainchild – the one that they built with blood, sweat and tears. They could have tribal knowledge about the current homegrown solution that ensures job security. When others move to a *aaS solution, they would no longer be single points of failure. In this case, there's conflict of interest that introduces bias in the vendor approval process.

- A committee of DevSecOps "experts" could be deployed to make *aaS selection decisions, which could potentially burn hundreds (sometimes thousands) of dollars per hour. Ironically, the total cost for these "analysis paralysis" weeks could outweigh the subscription cost of the new tool for the whole quarter. Which means, we could have signed a 3-month contract with this vendor to complete a PoC (proof of concept) that would have conclusively proved if our use cases are satisfied or not. The committee contributes on paper, while the PoC would have been working software with measurable RoI.

- DevSecOps "experts" may or may not have the right *aaS expertise at the right time, given that DevSecOps is breaking new ground in SaaS, IaaS and PaaS almost every quarter and the experts are learning as they go. Also, *aaS solutions have evolved at a breakneck speed and experts are being rendered ineffective by regular infiltration of new generations of concepts and tools. So, an expert is a great person to have around, however, to rely blindly without questioning and self-learning could be a bad idea that would inflate cost.

KEY TAKEAWAYS FROM CHAPTER 8

- A myth is only good until it's busted. Do everyone a favor - bust it.

- A working process that's in place for a number of years doesn't have to be right. Ask questions till you are satisfied. Don't get bogged down by fear of retaliation from "higher powers".

- Challenging status quo doesn't make someone a bad team player. If they are meant to stand out, they don't need to blend in. Welcome outliers to your team.

Chapter 9 | Resources

Whether we are doing Continuous Integration, Continuous Testing, Continuous Delivery or Continuous Deployment, the following resources help to design and build towards the Continuous paradigm.

Websites

1) Continuity's home: http://continuity.world/

2) Martin Fowler's home: https://martinfowler.com/

3) Jez Humble's home: https://continuousdelivery.com/

4) The Twelve-Factor App: https://12factor.net/

5) DevOps Topologies: http://web.devopstopologies.com/

Books

1) Continuous Delivery: Reliable Software Releases through Build, Test, and Deployment Automation, by Jez Humble and David Farley

2) Domain-Driven Design Reference: Definitions and Pattern Summaries, by Eric Evans

3) Continuous Delivery Pipeline - Where Does It Choke?: Release Quality Products Frequently And Predictably, by Juni Mukherjee

The Beginning Of A New Chapter In Your Journey

Cheers And Thank You!

Web: http://continuity.world

www.ingramcontent.com/pod-product-compliance
Lightning Source LLC
Chambersburg PA
CBHW041930240526
45473CB00034B/723